Climate and the Environment
The Atmospheric Impact on Man

JOHN F. GRIFFITHS, M.Sc., A.K.C., D.I.C.,
F.R.Geog.S., F.R.Met.S.
College of Geosciences, Texas A&M University

Paul Elek   London

© Elek Books Ltd., 1976

First published in Great Britain in 1976
by Elek Books Ltd.,
54–58 Caledonian Road, London N1 9RN

ISBN 0 236 40022 3

Printed in Great Britain by
The Camelot Press Ltd, Southampton

Contents

To Muz and Bads

Preface

Man can exist without water for only a day in the hottest deserts—but without air his time is measured in minutes. The air that reaches us has certain characteristics of temperature, humidity, and wind speed, among others, so it is really desirable that we should know something of these and the manner in which we react to them, both directly and indirectly. The applied meteorologist has long been aware of the fundamental role of the atmospheric impact on man, but during the past few years this realisation has come to many other people. The impetus has perhaps grown from times of adversity, the need to conserve energy, the necessity of producing more food and the possibility that our climate, even if it is not changing, may be growing more variable.

Those realisations have made us much more interested in learning about weather and climate, especially as they relate to our own lives. Generally, people want to know more about the weather of the next day or two—not just the cursory, non-committal 'tomorrow may be cool and rainy' given on some radio and television programmes (especially in Great Britain), but enough so that they can plan their activities. This book seeks to guide in the planning, by showing how we are really creatures of the atmospheric behaviour, that so many of our activities are influenced to a degree by the weather and climate. My hope is that this introductory type of text may be the cause of stimulating some readers to pursue the challenging science of meteorology in greater depth.

My thanks are due to Jack Grant and Roger Maynard for their drafting help, to Marilyn Kocurek, Lynn Scoggins and Dorothy Lorenz for secretarial assistance, and to my wife, Joan, for her great help with indexing and for her encouragement and understanding during the preparation of 'one more book'.

1

The Factors of Climate

1.1 Historical Survey

Mankind has always had to bow before the forces of Nature. In the earliest times Man, living without fire and clothing, could exist only in certain benign areas, but gradually, as his ability to modify his habitat grew, he could move into new areas and tolerate more severe conditions. Today we can live in almost any part of the world (and even on the Moon), if sufficient energy in the right form is available to us; but still, when Nature's biggest punches are thrown—hurricanes, tornadoes, tidal waves, earthquakes, drought or floods, for example —Man is rendered almost helpless in the face of such concentrated energy. Recently, it has become common knowledge that nothing is obtained without the expenditure of energy, and conservation is now the theme. In the light of this 'new' realisation, it is essential and interesting to appreciate the role played by the atmospheric component of man's environment, for, hopefully, we can learn to use what is given to us more efficiently.

The study of the atmosphere (*Meteorology*) is the common thread in this book. Meteorologists, aware that atmospheric phenomena are taking place over all scales of time, have seen fit to consider two main temporal divisions—weather (phenomena considered over short periods, say up to a few days) and climate (the synthesis of weather, or longer term manifestations). It is necessary to appreciate that there are scales of distance, area and volume, as well as a scale of time. Dimensions are referred to as macro- (a linear dimension of above about 100 km), meso- (from 1 to 100 km), and micro- (less than 1 km). These values are given here for guidance; no hard and fast divisions are, or ever can be, laid down.

Meteorology, especially through climatology, is a most important science that not only integrates physics and mathematics but teaches details of one of our natural resources, gives us facts concerning the part of the world in which we live (the *biosphere*), helps us understand a little better the problem faced in other areas and, as a bonus, allows us something of an armchair travelogue.

However, we must not think all can be solved readily. As Landsberg (1957) wrote,

The climatic conditions can be derived from a study of general atmospheric conditions and their local modifications.the circulation in the atmosphere is determined by a multitude of processes converting energies in the atmosphere. Some of these processes are well known, others barely studied. Their influence, interrelation and interaction are very complex. Therefore, no analytical and quantitative treatment of all the causes underlying climate can be given at the present time.

The problem revolves around the understanding of energy balance and fluxes. Many appreciate the impact of energy upon themselves but, as Oort (1970) has written,

> . . . What is less familiar is the central function of the atmosphere and the oceans in redistributing the incoming solar energy and hence in determining the "macroclimate" of the earth.
>
> Present forms of life could not endure the harsh climate that would exist if conditions of radiative equilibrium were to prevail at all latitudes (that is, if the incoming solar radiation to a zone were exactly balanced by the outgoing terrestrial radiation from that zone). In spite of certain long-term climatic changes, climatological records do not show an appreciable net heating or cooling of the earth and its atmosphere.
>
> The properties of the surface determine the thickness of the layer over which the available heat is distributed. In the case of an ocean surface, wave motions are quite effective in distributing the heat through a thick layer, sometimes extending down to a depth of 100 metres.
>
> The oceans and the atmosphere are strongly coupled systems and cannot very well be treated separately. The final circulation pattern is determined by the interaction of the two systems, each system influencing the other in a complicated cycle of events.

Although astronomy was perhaps the earliest science, climatology must have run it a close second, even if only covertly, for the hunting and nomadic peoples needed to be aware of seasonal rainfall patterns if they were to benefit from good grazing for wild or domestic animals. For agriculturalists, this knowledge is even more essential, as the Egyptian studies of Nile floods testify. In the first written messages of the Sumerians there are references to weather observations.

The first coordinated attempts, of which we are aware, to study climate were undertaken by the Greeks—from whom the word climate (*klima*, a slope) is derived. In fact, many of the words used in meteorology are obtained via the Greek language: the tropics—*trope*, a turning (of the sun's apparent path); Arctic—*Arctos*, the 'Great Bear' constellation; and even meteorology itself—*meteoros*, elevated. The earliest climatic classification was really a geographical one with the torrid zone (between the tropics), where there was no winter, the temperate zone (poleward of the torrid), and the frigid zone (north of the Arctic Circle), where some days the sun never appeared and no real summer existed. This system, linked to day length and solar position, is called 'Solar Climate'; its relationship with solar radiation is obvious.

Then, for many hundreds of years, there was an apparent hiatus in climatology, until in the twelfth century the Arab scientist, Idrisi, used a classification dividing the known world into seven zones, each with ten different regions. However, it is really in the last hundred and fifty years that the science has become developed, with such names as Alexander von Humboldt and Heinrich Dove leading the way and Lambert Quetelet advancing statistical ideas to collate and examine the mass of accumulating data (Landsberg, 1964). In 1873, Julius Hann and Wladimir Köppen were present at the first International Meteorological Organisation meeting in Vienna. These two men were to have a profound influence on climatology for many years by their example in ordering and classifying climates. Direct applications of meteorology to other disciplines were made by many individuals over the years, often during times of stress—war, or disasters such as locust outbreaks—but it was not until 1956 that the International Society of Biometeorology and Bioclimatology (as it was then called) was founded, thanks to the leadership of its Secretary, Dr S. W. Tromp of the Netherlands and the first President, Dr Fred Sargent, II.

1.2 Introduction

The climate of any region is not completely expressed by only one variable, such as temperature. Climate is a many-faceted phenomenon with complex interactions. In order to understand the complete picture, the climate is divided into a number of component parts called elements, of which, for example, one is temperature.

The science of climatology seeks to understand the variations in the elements by studying the determining causes, the factors, of climate. Once these are identified and their role clarified, then the patterns of climate become easier to comprehend, explain and forecast.

1.3   Primary Climatic Factors

The sun is the source of more than 99·97% of our total energy. This energy is the driving force behind atmospheric and oceanic movements. As Oort (1970) has remarked, 'The distribution of sunlight with latitude determines to a great extent the location of the major climate zones.' This concept is consistent, of course, with the Greek climatic system that was related to latitude and referred to as 'solar climate'. Clearly, then, the primary factor in climatology is *solar radiation.*

When solar radiation meets a medium, a part of the incident energy is reflected away, part is absorbed and the remainder is transmitted through the medium (Chapter 2). In climatological studies, it is the energy absorbed that is important, for it is this energy that heats the surface; of course, much of the reflected energy is later absorbed by other media and so also plays its role in the energy budget. Since the physical characteristics of the different media vary, the *receiving surface* (or *underlying surface*) must also be considered a basic factor.

In Table 1.1 the absorption percentages for various surfaces are given. For opaque materials, the sum of absorbed and reflected

TABLE 1.1

Absorption of solar radiation (%) by various surfaces

Water	90–95	Sand, wet	70–80
Road, black top	90–95	Desert	70–75
Forest, coniferous	85–95	Soil, light	55–75
Soil, dark	85–95	Sand, dry	55–65
Meadow, green	80–90	Human skin, blonde	55
Forest, deciduous	80–90	Snow, several days	
Human skin, dark	78–84	old	30–60
Concrete, dry	75–85	Snow, fresh	5–25

radiation will equal 100%, but for translucent objects (such as leaves) the total is expressed by absorbed plus reflected plus transmitted radiation. From this table we see that, for instance, dark soil absorbs more radiation than light soil and a black top road will absorb more than a dry, concrete road. The values in Table 1.1 have to be interpreted with care, however, for a complicating factor can arise. The more radiation a dry surface absorbs, the hotter it becomes, but if moisture is contained therein, some evaporation will occur and the temperature will be lowered. For this reason dry sand will reach a higher temperature than wet sand, although the latter absorbs more solar radiation. The resulting patterns are shown in Figure 1.1, the media in Group I being assumed dry, while some in Group II will have evaporation taking place.

It is seen, then, from Figure 1.1, that under identical radiative conditions, surfaces in section A will reach higher temperatures than those in section B. In addition, due to the conduction of heat, the air above A will become hotter than the air above B. Thus, the air will be less dense (hot air is lighter than cold air) and a pressure difference will be generated (Figure 1.1). When a pressure gradient is established, an air movement, attempting to reduce the difference, will begin, the air moving from a high pressure towards low pressure. This flow will not be a direct one because other aspects, such as the friction of the surface,

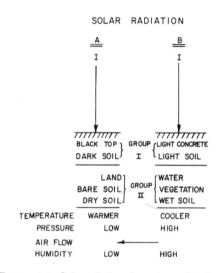

FIGURE 1.1 Solar radiation, absorption and its effects

the turning of the earth, and relief features will play a part. The friction is related to the nature of the receiving surface but two additional basic factors have now been introduced—*the rotation of the earth* and *topography*.

Radiation and the receiving surface have an effect on yet another climatic element—atmospheric moisture. If the amount of moisture in the air remains constant (fixed absolute humidity), an increase in air temperature will result in a lowering of the relative humidity. In addition, the radiative energy falling on surfaces B, Group II (Figure 1.1), will cause evaporation and an increase in the amount of moisture in the air. Thus, the air over surface B will, in general, have a greater relative and absolute humidity than the air over surface A. This change in humidity will have many ramifications because humidity affects cloud, sunshine, fog and precipitation. We see, therefore, that many of the climatic elements are already accounted for.

The next step may be a little more difficult to visualise. So far, we have thought in terms of air movement (winds), but another means of energy flow is by water movement (ocean currents). These transporters of heat and cold, whose importance will be discussed later, follow certain paths across the oceans, which are determined by rotation, topography (of the ocean floor) and another basic factor—*land-sea configuration*. Naturally, the distribution of the surfaces described in Figure 1.1 is also affected by this.

1.4   Land and Sea Patterns

We are all well acquainted with the pattern of the continents and oceans, a pattern in which land accounts for nearly 150×10^6 km², or

TABLE 1.2

Distribution of water and land between parallels (Sverdrup, 1952)

Latitude (°)	Northern hemisphere				Southern hemisphere			
	Water (10^6 km²)	Land (10^6 km²)	Water (%)	Land (%)	Water (10^6 km²)	Land (10^6 km²)	Water (%)	Land (%)
90–85	0·98	—	100·0	—	—	0·98	—	100·0
85–80	2·55	0·38	85·2	14·8	—	2·93	—	100·0
80–75	3·74	1·11	77·1	22·9	0·52	4·33	10·7	89·3
75–70	4·41	2·33	65·5	34·5	2·60	4·14	38·6	61·4
70–65	2·46	6·12	28·7	71·3	6·82	1·76	79·5	20·5
65–60	3·12	7·21	31·2	69·8	10·30	0·03	99·7	0·3
60–55	5·40	6·61	45·0	55·0	12·01	0·01	99·9	0·1
55–50	5·53	8·07	40·7	59·3	13·39	0·21	98·5	1·5
50–45	6·61	8·46	43·8	56·2	14·69	0·38	97·5	2·5
45–40	8·41	8·02	51·2	48·8	15·83	0·59	96·4	3·6
40–35	10·03	7·63	56·8	43·2	16·48	1·17	93·4	6·6
35–30	10·81	7·94	57·7	42·3	15·78	2·97	84·2	15·8
30–25	11·75	7·95	59·6	40·4	15·44	4·26	78·4	21·6
25–20	13·35	7·15	65·2	34·8	15·45	5·05	75·4	24·6
20–15	14·98	6·16	70·8	29·2	16·15	5·00	76·4	23·6
15–10	16·55	5·08	76·5	23·5	17·21	4·42	79·6	20·4
10– 5	16·63	5·33	75·7	24·3	16·90	5·06	76·9	23·1
5– 0	17·39	4·74	78·6	21·4	16·79	5·33	75·9	24·1
90– 0°	154·70	100·28	60·7	39·3	206·36	48·61	80·9	19·1

All oceans and seas	$361·059 \times 10^6$ km², 70·8%
All land	$148·892 \times 10^6$ km², 29·2%

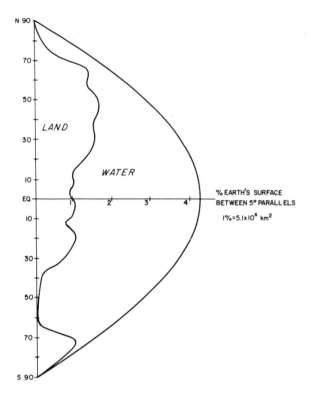

FIGURE 1.2   The pattern of land and water distribution

29% of the global surface, and water covers 361×10^6 km^2. In the southern hemisphere 81%, or 206×10^6 km^2, is water, while in the northern hemisphere the total is 61%.

In Table 1.2 a complete division of the distribution of water and land between 5° parallels is given and a graphical presentation is made in Figure 1.2. Note that there is more than twice as much land in the northern as in the southern hemisphere.

References

Landsberg, H. E. (1957) 'Review of climatology, 1951–55', *Meteorol. Res. Rev.*, American Meteorological Society, **3** (12), July, 1–43
Landsberg, H. E. (1964) 'Roots of modern climatology', *J. Wash. Acad. Sci.*, **54**, 130–41

Oort, A. H. (1970) 'The energy cycle of the Earth', *Sci. Am.*, **223**(3), 54–63

Sverdrup, H. N., Johnson, M. W. and Fleming, R. H. (1952) *The Oceans*, Prentice-Hall, New York, 1087 pp.

Suggested Reading

Critchfield, J. J. (1961) *General Climatology*, Prentice-Hall, New Jersey, 465 pp.

Miller, A. A. (1961) *Climatology*, Methuen, London, 320 pp.

2

Radiation and the Energy Budget

2.1 Introduction

The sun's output of energy, which for most practical purposes can be considered to remain constant, is about 56×10^{26} calories per minute or 29×10^{32} calories per year. At the mean earth–sun distance (15×10^{12} cm or nearly 93×10^6 miles) this is equivalent to approximately $1 \cdot 94$ cal cm^{-2} min^{-1} or $1 \cdot 94$ Langleys per minute (ly min^{-1}), an amount referred to as the solar constant. The exact value is not known, but as Gates (1973) notes, 'It is one of the most important of all fundamental quantities known to man. All life on earth depends on the solar constant remaining more or less steady with time.' If the solar constant altered by 1%, the mean temperature of the earth, which is now about 10°C (50°F), would change by about $0 \cdot 6$°C ($1 \cdot 1$°F). The earth–atmosphere system intercepts only one part in 2×10^9 of the solar output, but this amounts to $3 \cdot 67 \times 10^{21}$ calories per day, or, if spread uniformly over the globe, about 263 kly per year at the poles. Table 2.1 compares the solar energy with other large and medium-scale energy sources.

2.2 Radiation at the Top of the Atmosphere

Due to the ellipticity of the earth's orbit, the daily receipt of radiation by the earth shows a variation of some $\pm 3 \cdot 5\%$ during the year, being greatest at perihelion (3 January) and least at aphelion (5 July). The daily amount of radiation falling on a surface at the top of the atmosphere is dependent upon the day of the year and the angle, i, at which the solar beam impinges on the surface—the amount varying with sin i, being zero when the beam is parallel to the surface and at maximum when it is at right angles ($i = 90°$). Since the angle of incidence for a horizontal surface is dependent upon only two parameters, the latitude, l, and time of the year, it is possible to represent the variation in a single graph.

Figure 2.1 gives an idea of how direct radiation varies with the month for chosen latitudes. The important aspect to note is the small seasonal variation at the equator compared with that in temperate latitudes, a fact that helps lead to less variation in weather, especially temperature.

When the solar beam reaches the atmosphere, it is affected by the

TABLE 2.1

Large-scale energy sources (Sellers, 1965).

(Rates are relative to solar energy available—263 kly per year.)

One-fourth solar constant	1
Heat flux from the earth's interior	18×10^{-5}
Sun's radiation reflected from the full moon	3×10^{-5}
Combustion of coal, oil, and gas in the United States	7×10^{-6}
Energy dissipated in lightning discharges	6×10^{-7}
Energy generated by lunar tidal forces in the atmosphere	3×10^{-8}

Total energy of various individual phenomena and localised processes in the atmosphere. (Rates are relative to the solar energy intercepted by the earth—$3 \cdot 67 \times 10^{21}$ cal per day.)

Solar energy received per day	1
Melting of average winter snow during the spring season	10^{-1}
Monsoon circulation	10^{-2}
World use of energy in 1950	10^{-2}
Average cyclone	10^{-3}
Average hurricane	10^{-4}
Kinetic energy of the general circulation	10^{-5}
Average summer thunderstorm	10^{-8}
Detonation of Nagasaki bomb in August, 1945	10^{-8}
Burning of 7000 tons of coal	10^{-8}
Average local shower	10^{-10}
Average tornado	10^{-11}
Average lightning stroke	10^{-13}
Individual gust near the earth's surface	10^{-17}

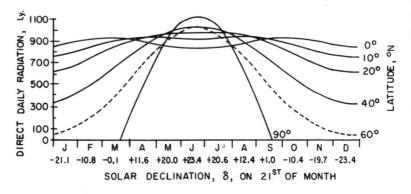

FIGURE 2.1   Annual variation in daily global radiation on a horizontal surface for chosen latitudes (Reprinted with permission from *Environmental Measurement and Interpretation* by R. B. Platt and J. F. Griffiths, 1972, original edition Litton Educational Publishing Co. Inc., reprinted by Robert E. Kreiger Publishing Co. Inc., 1972)

components of the atmosphere, being absorbed, reflected, and scattered. On average about 35–40% is reflected back to space, 45–50% reaches the earth's surface, and about 20–50% is absorbed in the atmosphere (Figure 2.2). However, these percentages vary greatly with location and season; for example, on a day of complete, thick

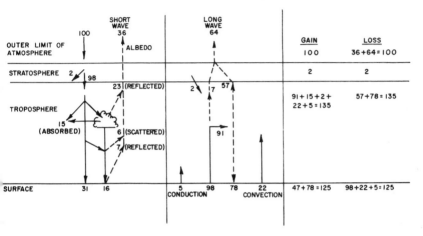

FIGURE 2.2   The balance of the energy budget of the atmosphere (from *Atmosphere, Weather and Climate* by R. G. Barry and R. J. Chorley. Copyright © 1970 by R. G. Barry and R. J. Chorley. Reprinted by permission of Holt, Rinehart and Winston, New York)

cloud cover, only about 5–10% may reach the earth's surface at that location. As the direct solar beam traverses the atmosphere, it becomes depleted, while an increase in the amount of reflected or diffuse radiation occurs. The solar radiation recorded at the earth's surface on a horizontal plane (the global radiation) then consists of two portions, the direct (sun) radiation and the diffuse (sky and cloud) radiation. There are two important differences between these components, (1) the direct beam is unidirectional, while the diffuse is omnidirectional (received from all the hemisphere of the sky, but *not* uniformly from all angles), and (2) the quality of the radiation differs between sun and sky (Figure 2.5). On cloudless days, the ratio of direct to diffuse is almost zero at sunrise and sunset, but increases to about 1·5 with the sun 10° above the horizon and to about 4 when the sun is at 40° altitude or greater.

The sun's altitude (α) can easily be calculated from

$$\sin \alpha = \cos l \cos \delta \cos H + \sin l \sin \delta \qquad (1)$$

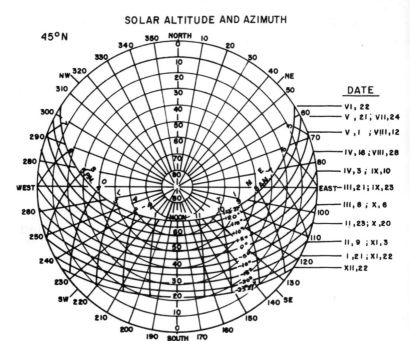

FIGURE 2.3   Solar altitude and azimuth for 45°N (Reprinted with permission of Smithsonian Institution Press from *Smithsonian Meteorological Tables*, Vol. 114 sixth revised edition, Robert J. List, 1966)

where l = latitude of station, δ = solar declination (*see* Figure 2.3), and H = hour angle (measured from the south).

The hour angle at 1.00 p.m. would be 15° and at 10.00 a.m. 330° ($= 360 - 2 \times 15$), but times must be corrected for the deviation of the longitude of the station from the local time standard, with a small correction (between $+14$ min and -16 min) for the irregular, apparent path of the sun through the heavens. The azimuth (A) of the sun can be obtained from

$$\sin A = \cos \delta \sin H \sec \alpha \qquad (2)$$

For sunrise and sunset $\delta = 0$, and $H = H_0$, where

$$\cos H_0 = -\tan l \tan \delta \qquad (3)$$

Equation (3) shows that day length is 12 hours ($H_0 = 90° = 6 \times 15°$) when $\delta = 0$, that is, at the equinoxes (21 March, 20 September) and when $l = 0$ (Equator), regardless of δ. Equations 1 and 2 are used to derive solar position diagrams, such as Figure 2.3 which gives the patterns for latitude 45°N. The patterns for other latitudes can be found in the *Smithsonian Meteorological Tables* (1966).

An important point to note is the great variation in day length with season as the latitude increases. This has great impact on both flora and fauna, for they become adapted to these seasonally occurring conditions. In Figure 2.4 the values of day length, civil twilight, and solar altitude at noon, for the summer and winter solstices, are given.

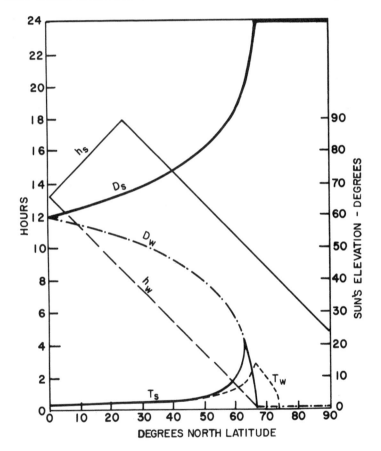

FIGURE 2.4 The length of daylight (D) and civil twilight (T), and the solar altitude at noon (d) at the summer (s) and winter (w) solstices (From *Principles of Climatology: A Manual in Earth Science* by Hans Neuberger and John Cahir. Copyright © 1969 by Holt, Rinehart and Winston, New York. Reprinted by permission of Holt, Rinehart and Winston)

On sloping surfaces, such as walls, roofs, and leaves, the equation of incidence of the sun's direct rays, i becomes quite complicated (*see* Section 2.6). However, for the simple case of a vertical wall

$$\cos i_v = \cos \alpha \cos (A-P).$$

where P is the direction (azimuth) in which the wall faces.

2.3   The Spectral Distribution of Radiation

Although it is general to use the concept of wavelength for radiation identification, for complete accuracy, wave number should be used. Wave number is proportional to the reciprocal of the wavelength and is independent of the medium through which the radiation is transmitted. The sun emits radiation in a broad band of wavelengths, but about 95% of its energy is contained within the range $0 \cdot 2$–$4 \cdot 0$ μm (μm = micron $= 10^{-4}$ cm). Before reaching the atmosphere, solar radiation comprises approximately 7% in the ultraviolet band ($< 0 \cdot 4$ μm), 46% in the visible ($0 \cdot 4$–$0 \cdot 7$ μm) and 47% in the infrared ($> 0 \cdot 7$ μm). In Figure 2.5 the type

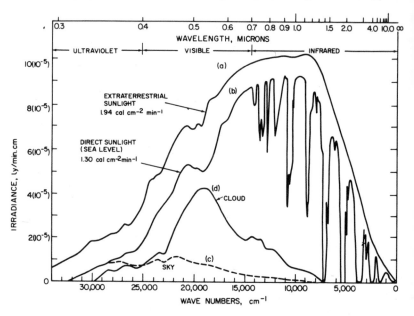

FIGURE 2.5   Spectral distribution of extraterrestrial solar radiation, of solar radiation at sea level for a clear day, of radiation from a complete overcast, and of radiation from a clear sky (after D. M. Gates, 1965)

of distribution (intensity v. wavelength, λ) is shown for various conditions, the upper line (a) being for solar radiation at the top of the atmosphere. This line fits very closely to a temperature (T)—dependent theoretical distribution (known as Planck's law), where intensity is proportional to $\lambda^{-5}/[\exp(c_2/\lambda T)—1]$, when $T = 6000$ K where K is degrees Kelvin and K = degrees Centigrade $+273.2$. However, at wavelengths below 0.5 μm, a better fit is $T = 4000$ K or 5000 K. The optimum value of T may change somewhat because, although in Section 2.1 it was stated that the sun's output was almost constant, there is some irregular, unpredictable variation in the ultraviolet part of the spectrum.

The sun's rays are absorbed selectively in the atmosphere, the ultraviolet by nitrogen, oxygen, and ozone, and the infrared by carbon dioxide and water vapour, so that, with the sun in the zenith ($\alpha = 90°$), the spectral distribution at sea level on a horizontal surface will appear somewhat like (b) in Figure 2.5. All wavelengths below 0.29 μm are removed, except at high elevations on mountains; this is fortunate because these extreme ultraviolet waves are injurious to most organisms.

As the sun's altitude decreases, the rays must traverse more of the atmosphere (optical air mass), so that depletion increases and the wavelengths of maximum intensity move towards the red end of the spectrum and the sun appears a deep red colour. Because blue light is scattered more than green, green more than yellow, and yellow more than red, the sky radiation is rich in blue—line (c), Figure 2.5. On cloudy days the water droplets absorb the infrared, while ultraviolet and visible wavelengths are scattered appreciably and give a whitish-grey look to the overcast sky—line (d), Figure 2.5.

A very important law in radiation is that all surfaces emit radiative energy proportionally to the fourth power of their surface temperature, on the Kelvin scale. If R is the amount of radiation in langleys per minute, then the Stefan-Boltzmann law states

$$R = \varepsilon \sigma T^4$$

where ε is the emissivity of the surface, and σ is a universal constant (Stefan-Boltzmann constant $= 8.13 \times 10^{-11}$ ly min^{-1}K^{-4} or 5.69×10^{-8} W m^{-2} K^{-4}).

For a 'black body', a surface which absorbs all incident radiation, $\varepsilon = 1$. In nature, objects with $\varepsilon = 1$ do not exist, although in some wavelengths the value may be close to unity. The absorptivity can, and does, vary with wavelength; for example, fresh snow may absorb only 5% of visible radiation but 95% of the infrared radiation.

Objects in the biosphere, such as animals, buildings, soil, vegetation, all have temperatures around 300 K (250–320 K) and their radiation is almost exclusively in the far infrared wavelength. This radiation, roughly from 4–50 μm is referred to as long wave (terrestrial or thermal) to distinguish it from the 0·2–4 μm band, the short wave or solar radiation. The wavelength of maximum intensity decreases as the temperature of the object increases; for example, the sun ($T = 6000$ K) has its maximum in the green-yellow (0·5 μm), while a stone at, say, 300 K will have a maximum energy output at 10 μm, in the far infrared. Most surfaces have an emissivity around 0·9 in the long wave radiation.

Earlier it was noted that about 45–50% of the short wave radiation was received at the earth's surface. In the same units, some 80 units are received as long wave radiation from the atmosphere, but about 100 units are radiated from the surface, while convective loss (latent heat) is about 20–25 units and conduction loss (sensible heat) is five units; leaving the surface in balance. A similar budget reckoning shows that the atmosphere is also in long period balance (Figure 2.2). It must be remembered that surfaces absorb the short wave radiation, become warmer, and then re-emit in the long wave.

Long wave radiation is absorbed by many of the atmospheric constitutents, especially carbon dioxide and water vapour. We have noted (Figure 2.5) that a cloudless sky is relatively transparent to short wave radiation, but this is not true for long wave radiation. However, in a band from about 8–13 μm there is a so-called 'window' through which energy is lost from earth to space (Chapter 5). This overall differential absorption of the atmosphere has a net effect of retaining energy in the earth-atmosphere system. This process is çalled the 'greenhouse effect', due to its similarity to conditions in a greenhouse in which the glass allows short wave radiation to pass, but prevents most long wave radiation from escaping. Since water vapour will absorb long wave radiation and then re-radiate it, much more terrestrial radiation will be returned to a radiating surface on a humid or cloudy night than on a clear and dry one. This fact, naturally, will prevent temperatures from falling as low on humid as on dry nights. Of course, any other cover, such as a tree or awnings, will lead to the same result.

2.4   World Distribution of Radiation

Figure 2.1 gives a *theoretical* pattern of radiation at the top of the atmosphere, the *actual* average annual global radiation at ground level being somewhat different (Budyko, 1974). The pattern over oceans is broadly related to latitude, but over land the distribution is affected greatly by cloud amounts, with the maximum in the sub-tropical deserts. The disposition of the short wave radiation by latitude is given in Figure 2.6. This shows the greater absorption at the surface in the

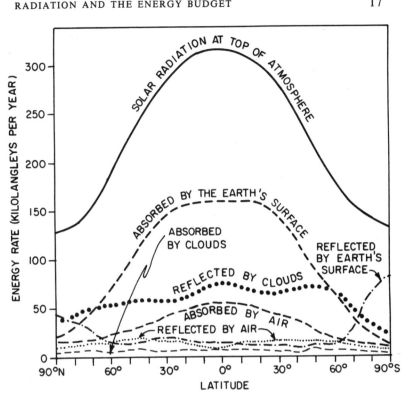

FIGURE 2.6 The average annual latitudinal disposition of solar radiation (in kilolangleys) (Reprinted with permission of University of Chicago Press from *Physical Climatology* by W. D. Sellers, 1965. Copyright © by the University of Chicago)

lower latitudes and the small amount of radiation used to heat the air at the surface in the polar regions.

Net radiation is the difference between downward short and long wave radiation and upward short and long wave radiation. Table 2.2 shows the ledger components of the radiation balance.

TABLE 2.2

The components of the radiation balance

Direction	Short wave radiation	Long wave radiation	All wave radiation
Downward	Direct and diffuse	Atmospheric	Total downward
Upward	Reflected	Terrestrial	Total upward
Sum	Net short wave	Net long wave	Net (all wave)

The world pattern of annual net radiation (Budyko, 1974) shows how the values increase fairly regularly outside the tropics but, owing to the large absorption of short wave radiation by water, there is almost twice as much net radiation over sea as in latitudinally corresponding land areas. Values are low over deserts because they experience a great net loss of terrestrial radiation. In Figure 2.7 we see the net radiation balance for the earth and atmosphere by latitude, and in Figure 2.8 the latitudinal variation in the components that show 'the central function of the atmosphere and the atmosphere and the oceans in redistributing the incoming solar energy'. Net radiation values are of particular importance in calculating evapotranspiration, although the component of advected or stored energy still remains unknown.

2.5   The Energy Balance

The ultimate goal of researchers in biometeorology is to understand the energy flows, or fluxes, for these dictate how the environment will act upon the organism or structure. In turn, the organism or structure will react to changes in these energy flows. The real aim of this book is to show how some of these interactions are related to man and his life.

The form of the balance equation is: response or adjustment of organism or structure = function (atmospheric variables, other abiotic variables, biotic conditions), and our attention is focused here upon the atmospheric variables. Because the problem is concerned with energy flows, the variables solar radiation, S, terrestrial radiation, T, convection, C, advection, A, conduction, P, and evaporation, E, will all play a part in determining whether a body gains or loses energy. In simple terms, the energy equation is

$$(S + T + C + P + A + LE)_{input} = (S + T + C + P + A + LE)_{output} + s \tag{4}$$

where s is a storage term and L is the latent heat of evaporation. Sometimes the radiation terms S_i, S_o, L_i, L_o are combined into a single term, R_N, the net radiation: $C_i, C_o, P_i, P_o, A_i, A_o$, are combined into H, a sensible heat transfer term: LE_i, LE_o became E a latent heat transfer term, and then (4) becomes

$$R_N + H + E + s = 0 \tag{5}$$

In the case of an organism, we must include terms to take care of photosynthesis, p, and the metabolic heat, M. Now (4) becomes

$$S(l - r - t - p) + \Delta T + H + E + s + M = 0 \tag{6}$$

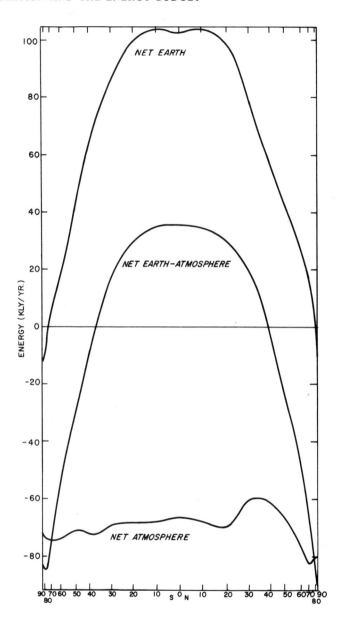

FIGURE 2.7 The average annual distribution of radiation balances. The scale is proportional to the global area at the latitudes

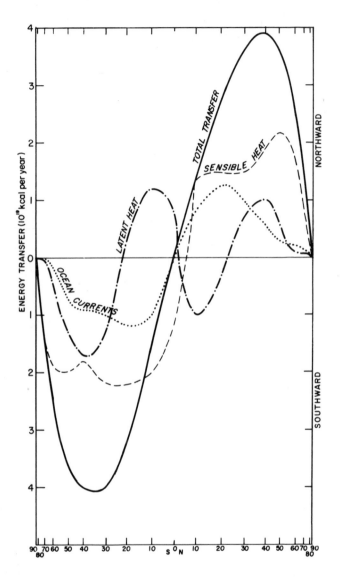

FIGURE 2.8   The average annual latitudinal distribution of the components of the poleward energy flux. The scale is proportional to the global area at the latitudes

where r is the short wave albedo of the receiving surface, t is its short wave transmissivity and ΔT is the net long wave radiation. It is clear from this that there exists a direct relationship between the heat and water components of the energy balance equations.

The water budget plays an important role in climatology due both to its implicit role in the energy balance equation and to its input at any one location through precipitation, evaporation, and runoff. The continuous movement of moisture in the land-air-water complex is referred to as the hydrologic cycle and a schematic picture of this is given in Figure 2.9. From this it is noted that only about 11% (12/106) of the precipitation over land derives from continental air masses, a

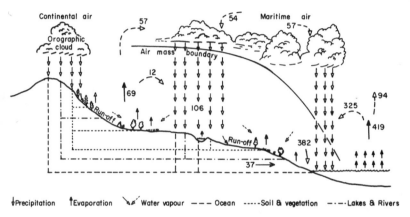

FIGURE 2.9   The hydrologic cycle with water mass movements (10^3 km³). (Reprinted with permission of Oxford University Press from *Applied Climatology: an Introduction* by John F. Griffiths, 1966)

percentage in keeping with extensive studies over large land masses (Benton *et al.*, 1950; Budyko, 1956). Naturally, those values vary greatly with geographical region and seasonally at any one location. The latitudinal variation of precipitation and evaporation is appreciable (Table 2.3). The water surplus around the equatorial belt stands out drastically when compared to the water deficits in the tropical and sub-tropical regions.

Lowry (1969) identifies four types of energy-budget systems. Type 1 is a two-dimensional interface; type 2 is a two-dimensional surface, such as a leaf, that is not an interface; type 3 is a three-dimensional interface, such as a plant canopy; type 4 is the three-dimensional system that is not an interface, such as an individual organism (man) or building (*see* Section 2.6).

TABLE 2.3
Variation of precipitation and evaporation with latitude (10^3 km³)

Latitude	Precipitation	Evaporation	Latitude	Precipitation	Evaporation
60°N	12·0	6·5	Equator	79·5	61·0
50°N	23·0	15·0	7°S	63·0	63·0
40°N	26·0	26·0	10°S	53·0	65·5
30°N	28·0	40·0	20°S	32·5	59·5
20°N	37·5	52·0	30°S	29·5	45·5
13°N	58·0	58·0	35°S	36·0	36·0
10°N	78·5	59·5	40°S	38·5	32·5
4°N	89·0	59·5	50°S	33·5	19·5
			60°S	14·5	8·0

2.6   Appendix

(1)   The general equation for calculating the incidence angle, i, between the sun's direct beam and a sloping surface is

$$\sin i = (\cos P \sin l \sin b + \cos l \cos b) \cos \delta \cos H + (\sin P \sin b)$$
$$\cos \delta \sin H + (\sin l \cos b - \cos P \cos l \sin b) \sin \delta$$
$$= \sin b \cos \alpha \cos (A - P) + \cos b \sin \alpha$$

where b is the angle of the slope and P is the azimuthal direction

(south $= 0$, and north $= 180°$)

This can be rewritten as

$$\sin i = \sin b \cos \alpha \cos (A - P) + \cos b \sin \alpha$$

where A is the azimuth of the sun (see Equation 2, page 12)

(2)   Lowry Energy Budget Systems

(a) *Type 1*
A two-dimensional interface using soil below the interface so that there is no radiant flux from below.

$$R_N + H + E + B = 0 \qquad\qquad (2.A)$$

where R_N is the net radiation, H is the sensible heat transfer by convection, conduction or advection, E is latent heat transfer and B is a grouping of sensible and latent heat streams for the lower hemisphere.

(b) *Type 2*

Two-dimensional systems that are not interfaces, such as a leaf, use an approach ignoring the mass of the leaf. The equation is then

$$a_s(S_d + S_u)_i + a_L(L_d + L_u)_i \\ + 2a_L\sigma T_L{}^4 + (H_u + H_d) + (E_u + E_d) = 0 \qquad (2.B)$$

where a_s is the absorptivity in the short wave, S is the short wave radiation, a_L is absorptivity in the long wave, L is the long wave radiation, H and E are as in (2.A), T_L is the leaf temperature, d and u represent downward and upward components, and i connotes inflow. The first two terms represent the radiant heat load on the system, while the last three terms relate to the radiative, convective-conductive, and latent heat dissipation.

(c) *Type 3*

This is the three-dimensional interface, such as a plant canopy. In this case, the storage and utilisation term, M, occurs and the equation becomes

$$(a_sS_{di} + a_LL_{di} + \varepsilon\sigma T_c{}^4 + H + E + B) + (H_h + E_h) + M = 0 \qquad (2.C)$$

where subscript h refers to horizontal components, and M has components comprising metabolic and photosynthetic energy, latent and sensible heat in the organism, latent and sensible heat in the canopy air. Note that the first group of six terms represents vertical fluxes while the next two are related to advection.

(d) *Type 4*

This system is three-dimensional, not an interface; type examples are an individual building or an individual organism. The equation then becomes

$$A_s(S_i + S_o) + A_L(L_i + L_o) \\ + A_H(H_i + H_o) + A_E(E_i + E_o) + M = 0 \qquad (2.D)$$

where i and o connote inflow and outflow components and the A's refer to the appropriate areas involved in the various processes.

References

Barry, R. G. and Chorley, R. J. (1970) *Atmosphere, Weather and Climate*, Holt, Rinehart and Winston, New York, 320 pp.

Benton, G. S., Blackburn, R. T. and Snead, V. O. (1950) 'The role of the atmosphere in the hydrologic cycle', *Trans. Am. Geophys. Union*, **31**(1), 61–73

Budyko, M. I. (1974) *Climate and Life*, Academic Press, New York, 508 pp.

Budyko, M. I. (1956) *The Heat Balance of the Earth's Surface*, Gidrometeoizdat., Leningrad (Trans. N. A. Stepanova, Office of Climatology, U.S. Weather Bureau, 1958), 255 pp.

Gates, D. M. (1965) 'Heat, radiant and sensible', *Meteorol. Monogr.* American Meteorological Society, 6 (28), 1–26

Gates, D. M. (1973) *Man and His Environment: Climate*, Harper and Row, New York, 175 pp.

Griffiths, J. F. (1968) *Applied Climatology*, Oxford University Press, Oxford, 118 pp.

Lowry, W. P. (1969) *Weather and Life*, Academic Press, New York, 305 pp.

Mather, J. R. (1974) *Climatology: Fundamentals and Applications*, McGraw-Hill, New York, 412 pp.

Neuberger, H. and Cahir, J. (1969) *Principles of Climatology*, Holt, Rinehart and Winston, New York, 178 pp.

Platt, R. B. and Griffiths, J. F. (1972) *Environmental Measurement and Interpretation*, Krieger, Huntington, New York, 235 pp.

Sellers, W. D. (1965) *Physical Climatology*, University of Chicago Press, Chicago, 272 pp.

Smithsonian Institution (1966) *Smithsonian Meteorological Tables*, 6th Edition, 527 pp.

3

Secondary Climatic Factors

3.1 Introduction

The basic factors described earlier (Chapter 1) actually act through a number of secondary factors to influence the weather and climate. To understand better the workings of the atmosphere, it is convenient to identify four of these subsidiary factors, namely, pressure systems, air currents, ocean currents, and air masses.

3.2 Pressure

It was described in Chapter 1 how the combination of radiation and receiving surface can give rise to appreciable air temperature differentials. Since hot air is less dense than cool air, a pressure difference also will result and this, in conjunction with differential heating in the atmosphere, helps to generate the average pressure patterns such as those shown in Figure 3.1.

It is seen in these diagrams how, during the summer, high pressures (anticyclones) are found over the ocean while lows (cyclones) form over land, due to the heating. In winter, extremely high pressures are located over central Asia, the northern United States and Canada because of the pronounced cooling of the large land masses. The pressures in Figure 3.1 are given in millibars, where 1 millibar is one-thousandth of a bar, a unit equal to 29·53 inches of mercury.

The subtropical high-pressure systems are the most prevailing of surface pressure features and change their positions little, especially in the southern hemisphere. There is a general thermal low pressure belt around the equator which is associated with a region of maximum insolation and moves with it. Figure 3.1 refers only to average pressure patterns and must not be confused with the systems of highs and lows that appear on the daily synoptic weather map. These maps generally show the transitory anticyclones and cyclones, ridges and depressions, that move across the broad pattern, each bringing its own weather characteristics.

Pressure in itself is seldom considered in a practical study of the environment, mainly because Man's responses to the small changes are

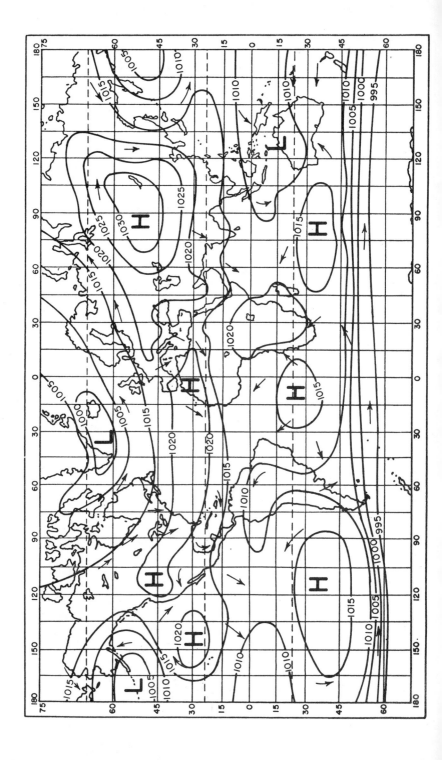

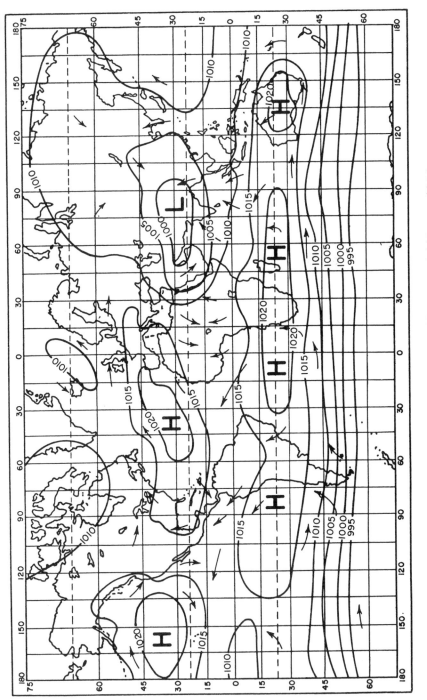

FIGURE 3.1   Average surface pressure and wind flow patterns in (a) January, (b) July

thought to be relatively unimportant. However, the pressure systems are fundamental in their effects on the air currents. If a pressure differential exists, the air will move in such a manner so as to reduce the difference—in other words, air will flow from a high pressure to a low pressure.

The closer the isobar spacing, the greater is the pressure-gradient force and the greater the wind speed. The air flow will not be direct between the pressure systems due to a deflective force caused by the earth's rotation, the coriolis force, which is proportional to the sine of the latitude and is, therefore, zero at the equator. The coriolis force causes a deflection to the right in the northern hemisphere and to the left in the southern hemisphere. In addition, the force due to friction between the air and the surface causes a change in both wind direction and speed. When all the forces are considered, the flow pattern becomes as illustrated in Figure 3.1, clockwise circulation around a high pressure and anticlockwise around a low pressure in the northern hemisphere with a reversal of this pattern in the southern hemisphere.

3.3   General Circulation

The major patterns of air flow throughout the year, or seasonally, are referred to as the general or primary circulation. In Figure 3.1 the wind directions for January and July are shown. This can be further simplified as given in Figure 3.2 where we see the effect of the maximum insolation (at the thermal equator) in forming a low pressure belt near the geographical equator. The vertical pattern of air flow, illustrated in the annulus around the globe, shows the ascending warm air at the equator. The thermal low leads to an in-flow of air into the 'doldrums' region, a flow that manifests as the north-east and south-east trade winds. In other words, the converging air leads to ascent while, similarly, diverging air leads to subsidence. The ascending air cools radiationally and begins to subside, becoming heated by compression, and arrives at the surface as hot, dry air—a major cause of the great subtropical deserts. Because there are only small variations in pressure patterns around the equator, the trade winds are generally remarkable for their constancy of direction (Figure 3.3a). In the temperate zones, where the westerlies predominate, the frequent movement of pressure systems reduces the constancy of direction (Figure 3.3b). The temperate zone patterns present an apparent wave motion around the poles, a motion known as the circum-polar vortex. It is large equatorward movements in this vortex that are believed to have caused some cooling or drying conditions recently experienced in parts of the world, although this is still unproven.

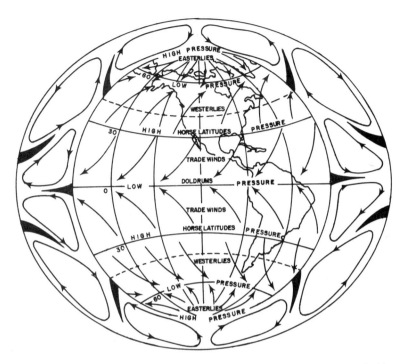

FIGURE 3.2 General circulation pattern—horizontal and vertical. (Reprinted with permission of Macmillan and Company from *Climatology and the World's Climates* by George R. Rumney. Copyright, George R. Rumney, 1968)

A little must be said about a circulation characteristic that is a seasonal feature, the monsoon circulation. The word derives from the Arabic '*mausim*' (season) and really is to be applied to pronounced wind direction changes of about 180° that bring a very different pattern of weather to an area. The 'classical' monsoon is found in India and China where it is associated, in its summer guise, with intense rainfall. However, other areas experience a similar wind reversal, a monsoon effect, and these include the southern states of the USA along the Gulf of Mexico, much of eastern Africa, and central-east Australia.

3.4 Ocean Currents

The atmosphere and the oceans are the basic media for the redistribution of solar energy and, hence, in the determination of the macroclimate of the earth. For a large water body wave motions are

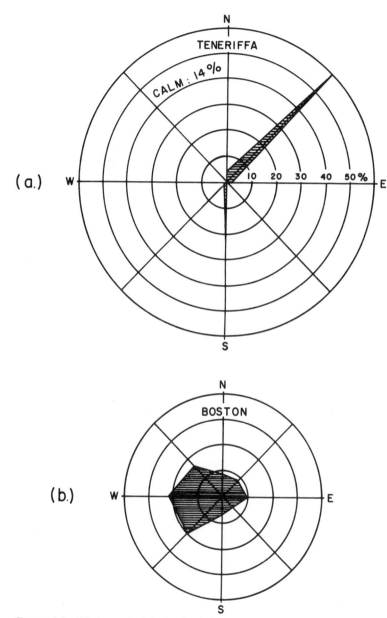

FIGURE 3.3   Wind roses for (a) a 'trade wind' and (b) a 'westerly' station. (Reprinted with permission from *Physical Climatology* by Helmut Landsberg, Gray Printing Company, 1960)

quite effective in distributing the heat through a thick layer, perhaps extending down 100 metres. Because the radiation is spread through such a depth, the daily variation in the temperature of the ocean surface is generally less than 1 deg C, whereas on land this variation may be as much as 50 deg C. Due to the effects of friction and the coriolis force, the average water flow is at right angles to the wind flow, a fact that has important consequences. For example, along the coasts of California and Peru the low level winds blow predominantly parallel to the coast and cause a water flow off-shore. As the surface water is swept away, deeper, and colder, water wells up to replace it, causing climatic effects upon the littoral. The pattern of the 'regular' ocean currents is shown in Figure 3.4. Warm currents are those bringing to the area, water masses that are relatively warm for that latitude, the reverse being true for the cold currents. The effects of these currents are shown in Figure 3.4; for instance, the warm winters that the Gulf Stream and the West Wind Drift (north Pacific current) bring to north western Europe and western North America respectively. For example, Tiree (Scotland, 56·5°N) has a mean annual temperature of 9·5°C (49°F) with a coldest month of 5·5°C (42°F), Copenhagen (Denmark, 56°N) a mean of 8°C (46·5°F) and a coldest month of 0·5°C (33°F) and Moscow (USSR, 56°N) a mean of 4°C (39°F) and a coldest month of −9·5°C (15°F)— illustrating the importance of the Gulf Stream in warming north western Europe. In Africa, the warm Agulhas current keeps Lourenço Marques (Mozambique, 26°S) only 2·5°C (4·5°F) cooler than Dar-es-Salam (Tanzania, 7°S) some 2100 km (1300 miles) closer to the equator, while making Vilanculos (Mozambique, 22°S) some 7°C (12·5°F) warmer than Walvis Bay (South-West Africa, 23°S) a region that is cooled by the northward-flowing Benguela current. The vast amount of energy involved in water movements is appreciated better when it is realised that the heat capacity of water is about five times that of soil and 3000 times that of air.

3.5 Air Masses

Through its nature as a source or sink of energy, the ocean acts as a great modifier of the air masses flowing over it. One large change the oceans cause is generally to bring the air immediately above them to a nearly saturated condition. When this modified, moisture-laden air reaches the land, the combination of topography and convection causes appreciable rainfall—witness the amounts received on Waialeale (Hawaii), Fernando Poo (Equatorial Guinea), and in Western Colombia and Cherrapunji (India), all in excess of 10 000 mm per year. Hurricanes and typhoons, spawned by the warm water of the

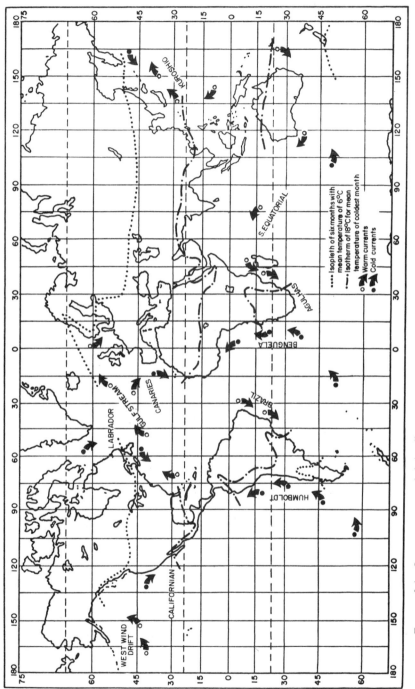

FIGURE 3.4 Ocean currents and their effects on temperature. (Reprinted with permission of Oxford University Press from *Applied Climatology: An Introduction* by John F. Griffiths, 1966)

subtropics, are special entities for the transport of energy in a poleward direction and play an important role in meridional movement.

Air masses, volumes of air having common characteristics in the horizontal, are developed in certain source regions (Figure 3.5). These are classified according to the source area—continental, c; maritime, m; tropical, T; polar, P; arctic, A—each type having its own peculiarities. As these masses move across the surface of the earth they are modified constantly so that, for example, a continental Polar (cP) air mass over central Canada is very different by the time it reaches the coast of the Gulf of Mexico. Nevertheless, it can still be identified as cP air by means of surface and upper air observations. Thus, air masses are determinants of the weather and the climate, but only in an intermediate role because their character has been moulded by the primary factors. The chief source regions for air masses are those large areas of uniform surface type which generally lie below semi-stationary pressure systems. Such regions are naturally to be found more over sea than land.

The boundaries between air masses are referred to as 'fronts'. Normally these are the narrow transitional or dividing areas between a warmer, lighter air mass and a colder, heavier air mass. The fronts of the temperate zones are generally of a turbulent type, as can be seen by the weather reports on television and the weather section in the newspaper. Basically, there are two types of fronts, warm fronts and cold fronts. If a warmer air mass advances over a wedge of cold air, we have a warm front—with a gentle slope (less than 1°). An early appearance of cirrus or cirrostratus clouds then leads to greater cloudiness, precipitation and increased humidity. This is usually the first sign of an advancing cyclone or low pressure centre. When the front passes there is decreasing cloudiness, and conditions characteristic of the warm air mass prevail.

The steeper cold front (cold air undercutting warm air) makes for more violence in the weather as the cold air displaces the warm air. The squall line (sudden showers and windshift) is a manifestation of the interaction between the cold air and the warm, moist, unstable air because conditions are changing more rapidly than through a warm front. Fronts can become either stationary, and thereby less intense, or occluded. An occluded front occurs when the faster-moving cold air overtakes the warm front and forces the warm air wedge aloft. Clouds and precipitation will occur in this warm sector.

In the tropical region of the world a major convergence region is identifiable, the Inter-Tropical Convergence Zone (I.T.C.Z.). The zone was originally referred to as a front but, since contrasts in air masses are not general in the tropics, the term zone is preferred. When located far from the equator the two air masses (mT and cT) can show sufficient

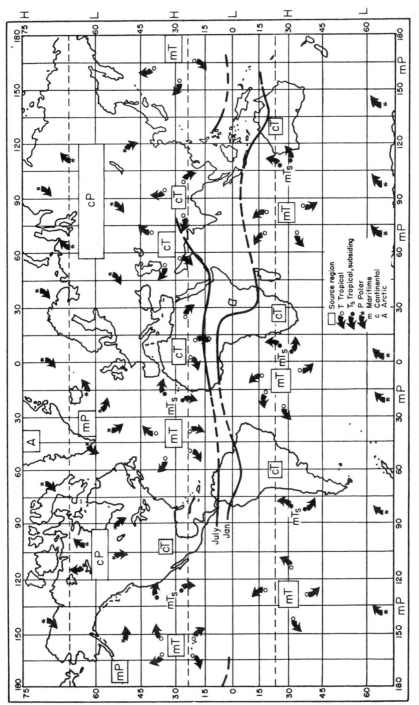

FIGURE 3.5 Air masses and source regions and average extreme positions of the I.T.C.Z. (Reprinted with permission of Oxford University Press from *Applied Climatology: An Introduction* by John F. Griffiths, 1966)

contrast for the old name Inter-Tropical Front (I.T.F.) to be used. In Figure 3.5 the average extreme positions of the I.T.C.Z. are shown. Apparently, rainfall is associated with the incidence of the I.T.C.Z. but the recently-realised fragmentary nature of the rainfall cells makes it appear that the tropical weather picture is much more complex than originally was thought.

References

Barry, R. G. and Chorley, R. J. (1970) *Atmosphere, Weather and Climate*, Holt, Rinehart and Winston, New York, 320 pp.

Griffiths, J. F. (1966) *Applied Climatology: An Introduction*, Oxford University Press, Oxford, 118 pp.

Landsberg, H. E. (1960) *Physical Climatology*, Gray Printing Company, Du Bois, Penn., 446 pp.

Rumney, G. R. (1968) *Climatology and the World's Climate*, Macmillan, London, 656 pp.

4

Climatic Patterns

4.1 Introduction

Each weather element has a certain degree of seasonality and persistence inherent in it. For example, temperatures are generally higher in summer than in winter (although specific days may not follow the trend) and rainfall today influences the probability of rainfall tomorrow. Because of this it is possible to identify certain 'preferred' patterns in climatic elements when the data from thousands of stations around the world have been studied. For instance, seasonal temperature curves in the northern hemisphere will tend to follow the patterns shown in Figure 4.1, London and Minneapolis being typical of the general

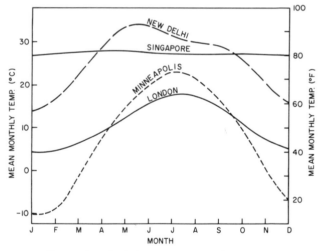

FIGURE 4.1   Typical seasonal patterns of temperature

curve, New Delhi showing the 'Gangetic' characteristic of a maximum before the onset of the monsoon rains, and Singapore depicting the uniformity of an equatorial station.

Precipitation is a little more complex as the reasons for precipitation have to be appreciated before the seasonal patterns can be interpreted. For precipitation to occur, moist air must be cooled and there are four main ways in which this can happen: (1) topographic ascent, (2) convectional or thermal ascent, (3) frontal ascent (*see* Section 3.5), and (4) convergence ascent (*see* Section 3.5). An analysis of seasonal patterns shows that, except for minor month-to-month fluctuations of 1 or 2% of the annual total, most stations exhibit a standard pattern of a dry month(s), after which the mean monthly totals increase monotonically to a maximum month(s), then decrease again monotonically. Of course, the pattern is not always as symmetrical as that shown by temperature and the wettest month(s) can occur at any time of the year.

The humidity of the air can be expressed in many ways. The actual amount of moisture in the air is called the actual vapour pressure, e. The maximum the air can hold at temperature, T, is the saturation vapour pressure, e_s, and Figure 4.2 shows how T and e_s are related; on average

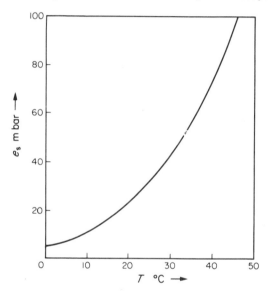

FIGURE 4.2 Saturation vapour pressure v. temperature

e_s doubles for an increase of 11 deg C (20 deg F) in T. The relative humidity is given by 100 e/e_s, the saturation deficit by $(e_s - e)$, and the dew point is that temperature at which condensation occurs when the air is cooled at constant pressure. The average daily pattern of temperature and humidity at a station (Figure 4.3) shows the great interrelationship between the two elements—a high relative humidity with a low temperature, and vice versa. The dew point is a much more conservative parameter and will vary little (± 2 deg C) during the day, unless an air mass change occurs.

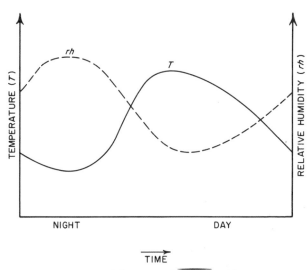

FIGURE 4.3   An average daily pattern of relative humidity and temperature

4.2 Classifications

No two places in the world have exactly the same climate but it is possible to describe areas having a fairly homogeneous climate. This technique of climatic classification has been very popular during the past hundred years and now over 100 classifications exist. However, as noted by Landsberg (1973) 'Much of this effort has been rather futile. The multiplicity of atmospheric parameters, their continuity in space, and their fluctuation in time do not permit a meaningful universal taxonomy.' Generally, a classification is derived so as to assist in a specific problem, such as examining the interrelationships between climate and vegetation or climate and human comfort. No one classification can be 100% efficient for all problems, but it should exhibit three important features:

1. coordinates the mass of data into a manageable form
2. is easy to apply
3. is based on meteorological principles

A fourth feature, suggested by Prohaska (1967), is that it should be directed towards limited, well-defined objectives based on atmospheric parameters, and limits germane to these objectives.

In order to have a quick grasp of the basically different, important climatic zones, a simple classification will be given using the threshold values shown in Table 4.1. Of the land area of the world 32·4% is in A, 23·3% in B, 16·2% in C, 14·8% in D and 13·3% in E. The deserts comprise 20·8% and zones with summertime maximum of rain cover 39·6%. About 11% of the land area is put into another thermal zone due to altitude and becomes a highland, H, zone. The largest regions are the long winter, summer rain (DS) with 13·3%, the cold, summer rain (ES) with 13·1%, the warm desert (BF) with 10·8% and the hot, short drought (A2) with 9·4%. Therefore, just these four zones comprise nearly half the world's land surface. This climatic zonation is shown in Figure 4.4. It must be noted that some potential zones, such as hot, extreme continental, do not exist, while others are too small to be of practical importance here.

There are many other climatic classifications, especially those due to Köppen and Thornthwaite, and the interested reader should refer to the book by Gentilli (1958).

TABLE 4.1

Classification of climates

TEMPERATURE		
Hot (A)	mean temperature of all months	$\geqslant 18°C$
Warm (B)	mean temperature of all months	$\geqslant 6°C$
Short Winter (C)	mean temperature of 7 to 11 months	$\geqslant 6°C$
Long Winter (D)	mean temperature of 3 to 6 months	$\geqslant 6°C$
Cold (E)	mean temperature of 0 to 2 months	$\geqslant 6°C$
Highland (H)	when altitude places station in a different thermal zone from what it would be if at sea level	
RAINFALL		
DESERT (F)	$R < (16 + 0·9T)$ R cm, $T°C$	
In A Zone—1	10–12 months each with $\geqslant 50$ cm	
2	7– 9 months each with $\geqslant 50$ cm	
3	3– 6 months each with $\geqslant 50$ cm	
In B, C, D, E zones	Uniform (U), $\dfrac{\text{driest consecutive 3 months}}{\text{wettest consecutive 3 months}} > 0·5$	

Spring   (V)—central month of wettest quarter in March, April or May
Summer (S)—central month of wettest quarter in June, July or August
Autumn (A)—central month of wettest quarter in September,
                          October or November
Winter (W)—central month of wettest quarter in December,
                          January or February
        (')—wettest quarter has 40–60% of the annual total
        (")—wettest quarter has over 60% of the annual total
Continental—annual temperature range > 17 deg C
Extreme
   continental—annual temperature range > 34 deg C

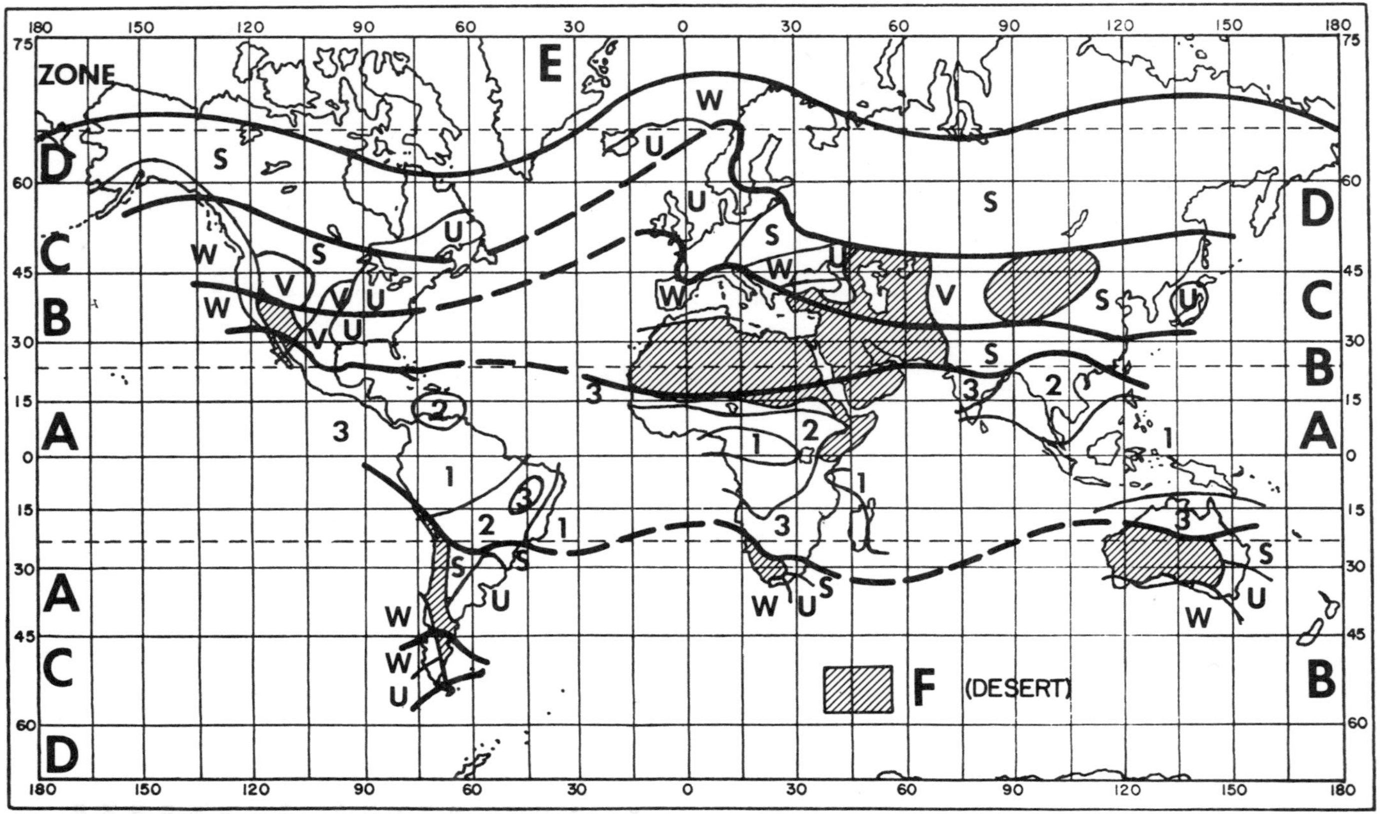

FIGURE 4.4 A simple climatic classification

4.3 Variability

Change is a fundamental characteristic of the atmosphere on all time scales. Investigators have shown how temperature varies greatly during a few seconds (Figure 4.5a) and during hundreds of thousands of years (Figure 4.5b). For convenience, climatologists often use a 30-year average to establish some 'normal' values, and trust that these are stable enough for most purposes. There is no reason to suppose that a 30-year mean of temperature or rainfall calculated for 1911–40 will be the same

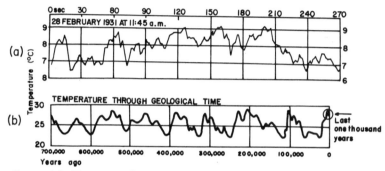

FIGURE 4.5   Temperature fluctuations over (a) seconds (b) hundreds of thousands of years

as one for 1941–70. It is well known that temperature changes from day to day but it must be realised that even mean annual temperatures can show large variation.

An interesting variation that occurs when climatic classifications are used is the shift from zone to zone that is experienced by most stations when yearly, as distinct from 'normal', conditions are used. For example, if only temperature is considered, so that a tropical zone is identified by a mean temperature of the coldest month of above 18°C (64°F), then, from Figure 4.6, it is seen that in 1940 none of Florida was tropical, in 1950 all was tropical, while in 1936, an average year, about 40% of the state was tropical.

Studies of rainfall have shown that the variation increases with decreasing amounts and is less variable in areas with more uniform rainfall. The world record rainfall amounts are shown in Figure 4.7 and an intriguing relationship is clear. The line shown is a very good fit to most points and enables an easy calculation of world maximum rainfall for any period to be assessed. The equation can be approximated by R (mm) $= 400 \sqrt{D}$, where D is in hours.

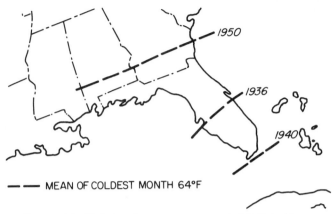

— — MEAN OF COLDEST MONTH 64°F

FIGURE 4.6   Variation of temperature patterns in S.E. United States.

Tables 4.2–4.8 list some of the world climatic records and give an idea of the ranges within which surface climatic patterns lie.

References

Gentilli, J. (1958) *A Geography of Climate*, University of W. Australia Press, Nedlands, 172 pp.

Landsberg, H. E. (1973) 'Developments and trends in climatological research', presented at Contemporary Celebration of the International Meteorological Organization/World Meteorological Organization, Vienna, Austria, September, 24 pp.

Paulhus, J. L. H. (1965) 'Indian Ocean and Taiwan rainfall set new records'. *Mon. Weather Rev.,* **93,** 331–335

Prohaska, F. (1967) 'Climatic classifications and their terminology', *Int. J. Biometeorol.,* **11**(1), 1–3

Suggested Reading

Critchfield, H. J. (1961) *General Climatology*, Prentice-Hall, New Jersey, 465 pp.

Holdridge, L. R. (1967) *Life Zone Ecology*, Tropical Science Center, San Jose, Costa Rica, 206 pp.

Thornthwaite, C. W. (1931) 'The climates of North America according to a new classification', *Geogr. Rev.,* **21,** 633–55

Trewartha, G. T. (1962) *The Earth's Problem Climates*, University of Wisconsin Press, 334 pp.

U.N.E.S.C.O. (1958) 'Climatology and microclimatology', Proc. of Canberra Symposium, *Arid Zone Res.,* XI, 355 pp.

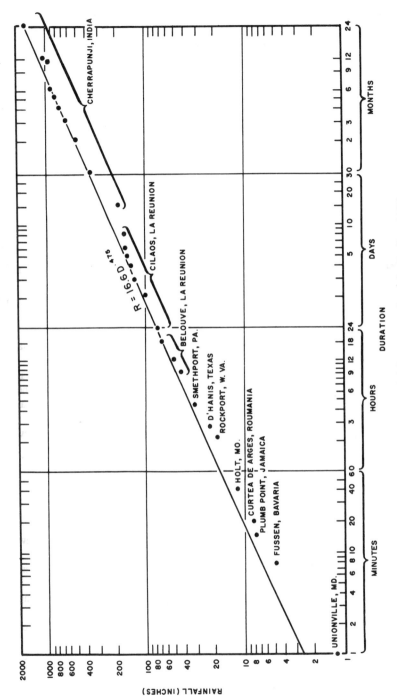

FIGURE 4.7 Maximum rainfall amounts v. time (after Paulhus, 1965)

TABLE 4.2
Some climatic extremes—temperature (°C)

	High		Low	
Absolute	58	Azizia, Libya (13 September 1922)	−88	Vostok, Antarctica (24 August 1960)
Mean monthly maximum	47	Bou-Bernous, Algeria (July)	−67	Vostok, Antarctica (August)
Mean monthly	39	Bou-Bernous, Algeria (July) Death Valley, California (July)	−72	Plateau Stn., Antarctica (August)
Mean monthly minimum	32	Dallol, Ethiopia (June)	−75	Vostok, Antarctica (August)
Mean annual	35	Dallol, Ethiopia	−58	Pole of Cold (78°S, 96°E)
Extreme range	107	(−70 to +37) Verkhoyansk, USSR	14	(13 to 27) Fernando de Noronha, Brazil (53 years)
Annual range	67	(−51 to +16) Oimyekon, USSR	0.5	Quito, Ecuador Marshall Islands
Mean annual diurnal range	21	(3 to 24) Bishop, California	4	(−1 to +3) Heard Island, Indian Ocean
Mean monthly diurnal range	28	(3 to 31) Quincy, California	2	(3 to 7) Macquarie Island, Indian Ocean
Extreme diurnal range	56	(7 to −49) Browning, Montana (23 to 24 January 1916)		South Pole (December)
Daily minimum			—	
Daily maximum	39	Death Valley, California	—	
Highest extreme minimum	21	Dallol, Ethiopia Pt. Moresby, New Guinea	−83	Vostok, Antarctica
Average of coldest month	31	Dallol, Ethiopia	—	
Average of warmest month			−33	Vostok, Antarctica (December)

TABLE 4.2 *(continued)*

Some climatic extremes—temperature (°C)

Spearfish, South Dakota —Temperature rose from − 20 to + 7 in 2 minutes
(07·32 — 22 January 1943)

Kansas City —Temperature went from 24 at 11.00 a.m. to − 10 at 6.00
6.00 p.m. and to − 12 at 11.00 p.m. (11 November 1911)

Rapid City, South Dakota—Temperatures on 22 January 1943

Time	05.30	09.40	10.30	10.45	11.30	16.00	19.30
Temperature	−21	12	−12	13	−12	13	−15

On this day, at one time, there was a temperature of 11°C at Lead, South Dakota, while at Deadwood (600 ft lower) 3 miles away, it was −27°C.

Hottest large cities

		Mean of hottest month	
Northern hemisphere	37	Jacobabad, Pakistan	(46–29)
	33	New Delhi, India	(41–27)
Southern hemisphere	28	Asuncion, Paraguay	(35–22)
	29	Manaus, Brazil	(33–26)

43 consecutive days with maximum over 49°C—Death Valley, California, 6 July—17 August 1917
162 consecutive days with maximum over 100°F—Marble Bar, W. Australia, 30 October 1923–7 April 1924

TABLE 4.3

Some climatic extremes—Precipitation (mm)

Highest annual total	26 467	Cherrapunji, India (1 August 1860–31 July 1861)
Highest annual means	11 684	Mt. Waialeale, Kauai, Hawaii (32 years)
	11 430	Cherrapunji, India (74 years)
	10 450	Ureka, Fernando Poo (7 years)
	10 300	Debundscha, Cameroons (32 years)
Lowest annual means	0·5	Wadi Halfa, Sudan (39 years)
	0·8	Arica, Chile (59 years)
		(Calama, Atacama Desert, Chile—rain never recorded. Iquique, Chile—no rain for 14 consecutive years)
Highest monthly total	9 296	Cherrapunji, India (July 1861)
Highest monthly mean	2 692	Cherrapunji, India (July)
Highest 5-day total	3 855	Cilaos, Reunion (March 1952)
	3 810	Cherrapunji, India (August 1841)
Highest 1-day total	1 870	Cilaos, Reunion (March 1952)
	1 245	Paishih, Taiwan (10–11 September 1963)
Highest 12-hour total	1 339	Belouve, La Reunion Island (28–29 February 1964)
Highest 4½-hour total	787	Smethport, Pennsylvania, USA (18 July 1942)
Highest 42-min total	305	Holt, Missouri, USA (22 June 1947)
Highest 20-min total	206	Curtea-de-Arges, Rumania, (7 July 1889)
Highest 1-min total	31	Unionville, Maryland, USA (4 July 1956)
Highest number rain days year average	325	Bahia Felix, Chile (4500 mm)
	311	Ponage, Pacific Island (4780 mm)
	310	Cedral, Costa Rica (3250 mm)
Highest number rain days in a year	359	Cedral, Costa Rica 1968
	348	Bahia Felix, Chile 1916
Highest annual total—snow	28 500	Paradise Ranger Station, Mt. Rainier, Washington, USA (1971–72)
Highest annual means—snow	14 783	Mt. Rainier, Washington
	14 605	Crater Lake, Oregon
Highest monthly total—snow	9 906	Tamarack, California
Highest 12-day total—snow	7 722	Norden Summit, California (1–12 February 1938)
Highest 6-day total—snow (single snow-storm)	4 420	Thompson Pass, USA (26–31 December 1955)
Highest daily total—snow	1 930	Silver Lake, Colorado (14–15 April 1921)
Greatest depth on ground	1 153	Tamarack, California (9 March 1911)

Cabo Raper, Chile, has 2210 mm of rain per year but has never had over 48 mm in one day (10 yrs)

Concepcion, Chile, has 1143 mm of rain per year but has never had over 20 mm in one day

Quseir, Egypt, has less than 2·5 mm of rain per year but received 33 mm in one day

Lobito, Angola, has 330 mm of rain per year but has received 536 mm in one day

Thrall, Texas, has 915 mm of rain per year but has received 970 mm in one day (September 1921)

TABLE 4.4

Some climatic extremes—radiation

Highest		
Mean annual	667 ly/day	La Quiaca, Argentine (3459 m)
	(85% of extraterrestrial radiation at 22°S)	
Mean monthly	955 ly/day	South Pole (2800 m) (December—8 years)
One hour	113 ly	Malange, Angola (1150 m)
	112 ly	Windhoek, S.W. Africa (1700 m)

TABLE 4.5

Some climatic extremes—sunshine

Highest			
Mean annual	4300 +	hours	Wadi Halfa
Continuous	60	hours	Antarctica, 9–12 December 1911
Day	24	hours	Antarctica, 9–12 December 1911
Monthly	14	hours/day	Syowa Base, 69°00′S, 39°35′E
Lowest			
Mean annual	500	hours	Laurie Island, 60°44′S, 44°44′W (44 years)
	550	hours	Argentine Island, 61°15′S, 64°15′W
Monthly		5% possible	Argentine Island (June)

TABLE 4.6

Some climatic extremes—pressure

Highest			
Measured	1081·8 mb	(31·91″)	Sedom, Israel (1275 ft b.m.s.l.) (21 February 1961)
Sea level	1084	mb (32·01″)	Agata, USSR (31 December 1968)
estimated	1079	mb (31·84″)	Barnaul, Siberia (23 January 1900)
Mean annual	1021·7 mb		Minusinsk, USSR
Mean monthly	1034·3 mb		Minusinsk, USSR; Troitsk Mine, USSR (January)
	1046	mb	Minusinsk, USSR (January 1954)

Perhaps Turfan Depression, Central Asia, has reached 1100 mb

Lowest			
Estimated	877	(25·90″)	Eye of typhoon 'Ida' 19°N, 135°E,
			60 miles N.W. Guam—24 September 1958
Measured	886·8	(26·29″)	Luzon Sea, on board Sapoerea, 18 August 1927

TABLE 4.7

Some climatic extremes—wind speed (surface)

Highest	mph	km h⁻¹	
	mph	*km h⁻¹*	
Absolute	231	372	Mt. Washington, New Hampshire, USA (1917 m), 12 April 1934
1 Day	108	174	Port Martin, Antarctica 21–22 March 1951
Annual average (USA)	35·4	57	Mt. Washington
Annual average (USA)	43	69	Cape Denison, Antarctica
Monthly average	65	105	Port Martin, Antarctica, March 1951

In tornadoes, speeds in excess of 500 mph (800 km h⁻¹) have been calculated.
Coast of Commonwealth Bay, Antarctica, has recorded a number of gusts in excess of 200 mph (320 km h⁻¹).

TABLE 4.8

Some climatic extremes—fog

Highest		
One year	7613 hours	Willapa, Washington, USA
Annual average	2552 hours	Cape Disappointment, Washington, USA
	1874 hours	S. Orkneys, Antarctica
	1580 hours	Moose Peak, Maine, USA
	120 days	Grand Banks, Newfoundland, Canada

5

Environmental Measurement

5.1 Introduction

It must be realised that instruments used in normal meteorological studies, of the standard type where the sensing elements are usually exposed at a selected height above the surface, are not necessarily those to be advised in the taking of environmental or micro-habitat type of measurements. Sometimes it is possible to utilise one piece of apparatus in both studies, but often the instrument has to be modified or redesigned. Because of this, the instruments described in *Meteorological Instruments* (1965) or *Handbook of Meteorological Instruments* (1965) are not as applicable to habitat studies as those given in *Environmental Measurement and Interpretation* (1972) and *Survey of Instruments for Micrometeorology* (1972).

Middleton and Spilhaus (1965) have suggested that an instrument should have four basic properties:

1. It should be as accurate as required for the problem
2. It should be as sensitive as needed to make full use of its accuracy
3. It should be robust and durable to give desired length of service
4. It should be convenient, simple and cheap as may be compatible with other requirements.

Many varieties of instruments are used in environmental studies and the interested reader should consult the bibliography of Griffiths and Griffiths (1966).

5.2 Radiation

Most radiation instruments use a method of conversion of heat energy via temperature change, expansion, or distillation in order to activate some element. There are only about 500–600 regularly reporting radiation stations in the world, and by far the greater number of these are concerned with the measurement of global (sun and sky) short wave radiation on a horizontal surface. A few other stations report, on a permanent basis, the measurement of direct solar radiation (made with a pyrheliometer) or, more recently, net radiation measurements, embodying both short and long wave radiation, measured by means of a balance pyrradiometer or a net radiometer. Glass can be used to cover and protect the elements when measuring short wave radiation because it is transparent in the $0 \cdot 3$–$4 \cdot 0$ μm band, but other materials such as polyethylene must be used if long wave or net measurements are required, since glass will absorb some long wave radiation. Radiation instruments are generally quite expensive and need good servicing. Instruments that cost, together with their recorders, approximately \$1000 can be expected to achieve an accuracy of approximately ± 3–5%, whilst the cheaper models may only achieve accuracies on the order of $\pm 10\%$, although this is likely to be improved when dealing with only daily totals.

Global radiation has two components one of which, the direct, is unidirectional whilst the other, the diffuse or sky radiation, is omnidirectional. This means that to calculate the short wave radiation falling on some surface it is necessary to have information concerning the relative magnitudes of these two components. The easiest way in which this may be achieved is by the use of a shadow ring. This is a metal ring placed at a distance of approximately 30 cm (1 ft) from the sensing element in such a way that it casts a shadow across the unit and prevents any direct radiation from being measured. The unfortunate aspect is that the shadow ring also cuts out a pronounced portion of the sky radiation, and empirical methods have to be derived for effecting a correction to the recorded amount. These errors may amount to as much as one-third of the diffuse radiation.

5.3   Radiation and Sunshine

Owing to the fact that radiation measurements are not made at a sufficient number of stations around the world, many investigators have attempted to use an empirical relationship between radiation and sunshine, the latter element having many more recording stations available. With this in mind, studies have been made of an empirical relationship that was first suggested by Angstrom and, in its present and most useful form, has been modified by Glover and McCulloch (1958). A form of this equation is given below

$$Q/Q_0 = 0 \cdot 29 \cos l + 0 \cdot 52 \, n/N$$

In this equation n/N is the ratio of actual sunshine to possible sunshine, while Q_0 is the short wave radiation received, during the chosen period, at latitude l, at the top of the atmosphere on a horizontal plane (Figure 2.1). This equation should only be used for calculations of long-term averages of radiation. Individual days, or even weeks, can stray greatly from such an equation. If possible, it is often better to derive one's own form of this equation for the climatological area with which one is dealing and to calculate the line of best fit and the correlation coefficient.

5.4   Temperature

Instruments for measuring the temperature, called thermometers, are generally one of four types: liquid-in-glass, liquid-in-metal, electrical (resistance or thermocouple type), or deformation (bimetallic strip). Usually when making a measurement of the temperature we are concerned more with the condition of the air or the soil than the temperature of the actual thermometer, which of course has physical characteristics very different from those of air, soil, water, leaf or animal. Therefore, it is usual, for meteorological purposes, to reduce the radiation error on these measuring elements by sheltering the thermometer. Generally this is not possible when the instrument has to be used for micro-environmental measurements and so, in order to reduce the radiation error, the measuring element must be made as small as possible. This has led to great use of resistance thermometers, especially thermistors, and thermocouples. If thermistors are to be used, then it is essential to have the correct type of element that will show maximum resistance change over the range of temperatures to be studied. Thermocouples have numerous sources of errors and reference should be made to Platt and Griffiths (1972) or Monteith (1972) for further details.

Recently the method using remote sensing, via infrared thermometry, has been improved. This method utilises radiation from a body, normally in the 8–13 μm range, and assumes a constant emissivity, usually about 0·95 to convert from incident radiation to temperature. If one is sensing a surface for which the emissivity is greatly different from the assumed value, corrections must be applied.

5.5   Moisture

5.5.1   Humidity

Psychrometers are the most common form of humidity measuring device and they have numerous errors; unfortunately, all of the errors tend to give too high a reading of the relative humidity and therefore are not compensatory. From a psychrometer, measurements of the dry and wet bulb are obtained and tables used for the calculation from these figures of absolute or relative humidity.

The hygroscopic technique, generally utilising human hair, is a most useful one under some conditions. It must be remembered that the hair only expands by about 2·5% in the range 0–100% relative humidity. At 20% r.h., 40% of the elongation has taken place, while at 80% r.h. 91% of the elongation has occurred. Unfortunately, the elongation is not a linear function of the relative humidity and corrections have to be made by the use of a cam, otherwise it is not possible to planimeter the recording to obtain a mean relative humidity.

A condensation technique necessitates the removal of a sample of the air and cooling this to the dew point. This is usually impracticable in most micro-environmental studies.

5.5.2 Precipitation

Rain gauges vary pronouncedly in size, from the Canadian one that covers 10 square inches (64 cm²) to the US model covering 50 square inches (323 cm²). The 8 inch (20 cm) diameter gauge covers only one eighty-millionth of a square mile and it is obvious that in all cases we are dealing with an extremely small sample—from which we are forced to make deductions. There are two classical works that are outstanding as surveys of precipitation and the instruments; these are by Kurtyka (1953) and Israelsen (1967). Israelsen's findings are that:

1. It is generally assumed, and erroneously so, that the gauge having the largest catch is the most accurate;
2. True areal precipitation remains unknown, so one cannot say that one gauge is better than another for indicating true precipitation. All that can be said is that the particular gauges used measure different things, or the same things differently.
3. Some investigators indicate that a gauge buried with its rim at ground level and surrounded by wire mesh or mat to prevent splashing will be the best.

Most of the rain gauges utilised (about 100 000 around the world) are of the direct measuring type, wherein one has to make a measurement of the rain at a specified time and interval—usually once per day. The number of recording type gauges in use is at least one order of magnitude less that the number of spot reading gauges. An accurate knowledge of rainfall amounts is important, for a fall of only 0·1 inch is equivalent to 10 tons of water per acre (1 mm = 9×10 kg ha^{-1}). A further complication arises due to the effect of wind speed on the catch of a gauge; at 4 m s^{-1} (9 mph) the deficit is about 10%, and increases to 50% at 12 m s^{-1} (27 mph). Since wind speed increases with height, the necessity for a standard exposure height is clear.

Snow measurements are extremely difficult to make. The usual technique, in regions with large snow fall, is for a measuring stick to be inserted and the accumulation of snow recorded by an observer. However, the stick itself can cause piling or scouring of the snow, for this is a case in which the measuring element affects the whole nature of the environment.

In spite of more than 2000 years' experience in measuring precipitation, Man has not yet perfected his instruments and techniques to the extent of his being able to obtain, routinely, precipitation measurements with the desired degree of accuracy. As Kurtyka concluded many years ago, so must we repeat today 'there remains much to be accomplished in measuring precipitation'.

5.6 Wind Flow

We must consider two aspects of air flow—the direction and the speed. Direction is normally measured by use of a vane, smoke drift, or even an eye estimate but, like the anemometer that is used for speed measurement, the instrument must be exposed away from obstructions that may cause local deviations in the flow. However, if one is attempting to measure the flow around bushes or buildings then it is absolutely necessary to expose the instrument in the position for which the measurements are required. Some types of anemometers utilise wind pressure in a form of windmill. Care must be taken then that the wind does not change direction so completely that the mill, in effect, winds and unwinds and finally records zero.

References

Glover, J. and McCulloch, J. S. G. (1958) 'The empirical relation between solar radiation and hours of sunshine', *Q.J.R. Meteorol. Soc.*, **84**, 172–5

Griffiths, J. F. and Griffiths, M. J. (1966) *A Bibliography of Meso- and Micro-Environmental Instrumentation*, U.S. Dept. Commerce, E.S.S.A., Tech. Note 43-EDS, 352 pp.

H.M.S.O. (1965) *Handbook of Meteorological Instruments*, Pt. 1, 458 pp.

Israelsen, C. E. (1967) *Reliability of Can-type Precipitation Gauge Measurements*, Utah Water Research Lab., 73 pp.

Kurtyka, J. C. (1953) *Precipitation Measurements Study*, Illinois Water Survey Division, Report of Investigation No. 20, 178 pp.

Middleton, W. E. K. and Spilhaus, A. F. (1965) *Meteorological Instruments*, University of Toronto Press, Toronto, 286 pp.

Monteith, J. L. (1972) *Survey of Instruments for Micrometeorology*, Blackwell Scientific Publications, London, I.B.P. Handbook No. 22, 263 pp.

Platt, R. B. and Griffiths, J. F. (1972) *Environmental Measurement and Interpretation*, Kreiger, New York, 235 pp.

6

Meso- and Micro-climates

6.1 Introduction

So far, we have been concerned with meteorological phenomena that influence a large area, tens of thousands of square kilometres or more. However, in applied aspects, it is often the immediate or local environment that is of greatest importance. For example, it is the conditions surrounding a plant (its micro-habitat) that influence its growth. It is essential, therefore, to study some of the smaller scale climates, referred to as meso-climates (a few square kilometres or more) and micro-climates (the smallest areas).

6.2 Meso-climates

There are four important meso-climate variables that are worthy of study—air flow, rainfall, temperature and radiation.

6.2.1 Winds

In Chapter 3 we examined the primary winds, those involving great energy transports. In this section the emphasis is on the secondary, smaller scale wind flows, namely land/sea breezes, gravity winds, mountain/valley winds, and föhn winds.

The land/sea breeze circulation is a phenomenon associated with radiation and the receiving surface (Figure 1.2). Solar radiation will heat a land surface more than the adjacent water so that a higher temperature and therefore a lower pressure will develop there. Such a situation will generate an inflow of air from the sea, a sea breeze of cooler, moist air that is usually relatively shallow (100 m or so). On overcast days, or if the adjacent land is heavily wooded or swampy, the land/sea temperature difference may take many hours to, or may never, reach a value sufficient to generate a sea breeze. The opposite situation occurs at night, when the sea cools less rapidly than the land and a circulation develops from land to sea (Figure 6.1). The speed of the land breeze is generally less than that of the sea breeze, and the depth of activity is less.

A gravity wind is a phenomenon associated with radiative cooling of

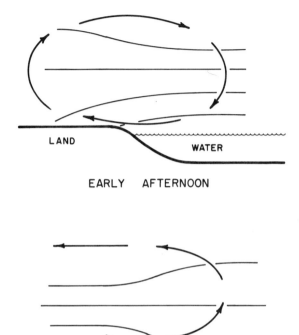

EARLY   AFTERNOON

LATE   EVENING

FIGURE 6.1   Land/sea breeze circulation

the surface. The overlying cooled air will begin to move due to the greater air density if the surface is sloping and a slow, cool flow will result. Such a wind is often called a katabatic (downslope) wind and needs a slope of only a few degrees to originate. It can be identified in cities along sloping roads. At times the cold air drainage can cause chilling of the air below dew point and fog pockets can result—a most dangerous situation. Such winds develop under clear skies and calm conditions, but are generally of slow speed.

The phenomenon of mountain/valley winds is depicted in Figure 6.2. The heated slopes induce an upslope (anabatic) flow during the daylight hours—the valley wind. At night, the gravity type flow drains into the valley as a mountain wind.

The föhn wind, called Chinook in the Rockies (Santa Ana in

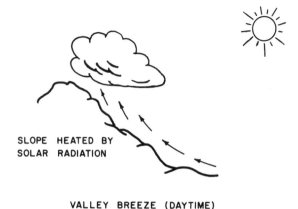

SLOPE HEATED BY
SOLAR RADIATION

VALLEY BREEZE (DAYTIME)

SLOPE COOLED BY
OUTGOING LONG WAVE
RADIATION

MOUNTAIN BREEZE (NIGHT)

FIGURE 6.2   Mountain/valley winds

California, Zonda in Argentine), occurs on the lee side of most mountain ranges. It is a hot, dry wind generated when humid air ascending the windward side loses most of its moisture by precipitation and descends on the lee side under dry adiabatic heating (Figure 6.3). A föhn wind can act almost like a warm front, and brings about very dramatic increases in temperature, the most outstanding being the change from −20°C (−4°F) to 7°C (45°F) in 2 minutes at Spearfish, S. Dakota, on 22 January, 1943.

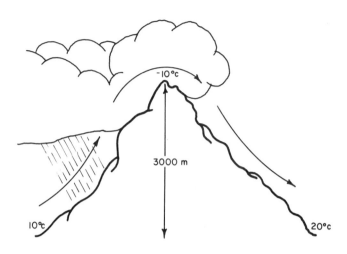

FIGURE 6.3   Chinook or föhn wind

It is logical to refer to hurricanes (named typhoons in the Pacific and cyclones in the Bay of Bengal) in this section as their real impact generally, at any instant, is on the meso-scale. Normally spawned over warm waters where surface temperatures exceed 27°C (80°F), these rotating storms can bring winds in excess of 240 km h^{-1} (150 mph) and torrential rains. However, the worst destruction to life and property is usually associated with the storm surge of water, and terrible disasters have occurred, such as the one million people killed in Bangladesh in 1970, or the 300 000 killed in Haiphong in 1881. On average, some 6–8 hurricanes affect the land areas of North America annually, the worst being in 1900 when Galveston, Texas, was completely destroyed and at least 6000 people died. Nowadays, due to satellite and aircraft tracking, an excellent warning system has reduced the loss of life appreciably, although the development of areas that are subject to hurricanes has led to great monetary increases in hurricane damage.

6.2.2   Radiation
Few areas have a sufficient density of pyranometers to make a meso-scale study possible, but a special investigation in Wisconsin has given some important clues. Even on clear days there was a large variation not accounted for by the latitude changes (from about 42–47°N) but by differences in the transmission coefficient. Daily changes in the standard deviation (a measure of variation) of values of the 17 stations ranged from 1–50% of the clear day global radiation.

6.2.3   Temperatures

The importance of small changes in temperature and the big changes they can induce in fundamental quantities is shown in Figure 6.4, from an article by Manley (1946). As will be seen, at location E about 2 months of freeze-free conditions exist, at C (only 3 deg C or 5 deg F warmer) this increases to over 4 months, and at A (about 6 deg C, 11 deg F warmer) this jumps to 7 months.

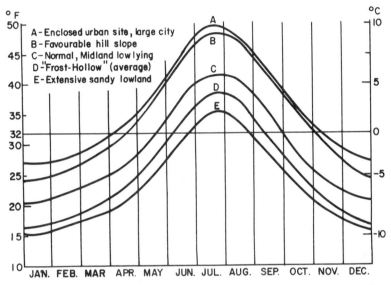

FIGURE 6.4   Seasonal temperature variation at selected sites (after Manley, 1946)

One of the most outstanding meso-scale temperature variations occurred around Rapid City, South Dakota, on 22 January 1943. Very cold Arctic air (temperatures around −18°C or 0°F) covered the region when minor pressure changes caused the air to undulate like waves on the sea. The thermograph record is perhaps the most unusual ever noted. At 5.30 a.m. the temperature in Rapid City was − 21°C (− 5°F). A west wind set in and at 9.40 a.m. the temperature was 12°C (+ 54°F). The wind subsided and at 10.30 a.m. the mercury stood at − 12°C (+ 11°F) rising to 13°C (+ 55°F) in a quarter of an hour as the west wind returned. By 11.30 a.m. the air was calm and the temperature stood at − 12°C (10°F), repeating this fluctuation twice more during the day. It is reported that at one time the temperature at Lead (South Dakota) was + 11°C (52°F) while at Deadwood, 200 m lower and 5 km away, it was $-27°C$ ($-16°F$). The rapid temperature changes caused plate glass windows in Lead to crack.

6.2.4   Rainfall

Rainfall can exhibit great variation over small distances, and an extreme case occurred in Texas in September 1921 when the greatest-ever 24-hour total of rain in the USA, 38·4 inches (975 mm) was measured at Thrall. The isohyetal pattern of the storm showed that a change from 5 to 35 inches (12·5 to 87·5 cm) occurred over a distance of only about 40 km (25 miles).

6.3   Micro-climate

6.3.1   Introduction

The micro-climate can be described as the climate in a small space. As Gates (1973) has suggested 'The climate surrounding an organism is the climate which is really significant for the comfort, behaviour and viability of the organism,' a sentiment I believe the astronauts on the moon, safe in their space suits, would endorse wholeheartedly.

Micro-climates were referred to by Theophrastus when, around 320 B.C., he wrote 'Hollow places, and such as are sheltered from external winds, are chilled by winds arising locally.' In the late eighteenth century Gilbert White investigated temperature differences over short distances and in 1893 Theodor Homen was studying the heat budget of various soils in Finland. The great work, and invaluable reference, on this subject is *The Climate Near the Ground* by Geiger (1965) and all who are fascinated by the subject of large climatic variations over small distances should study this publication.

One of the most comprehensive studies in micro-climates was made in the Neotoma Valley in Ohio during 1939 to 1944, where 109 stations were installed in an area of 0·6 km². A comparison for 1942 between measurements from this network and those from 88 stations of the macro-scale network covering the 113 000 km² of the whole state gave, for example, a range of frost-free period of 138–197 days in Ohio and 124–276 days in Neotoma (Geiger, 1965). It is clear that temperature variations on the micro-scale can be as great as would be experienced by changing climatic zones.

6.3.2   Radiation

Radiation over bare soil shows a maximum receipt at the surface, very little being absorbed by the air. However, in a grass stand there is gradual absorption until only a small percentage is received at the surface. This naturally influences the thermal patterns. A similar situation exists in a forest. Illumination (light) is also drastically reduced below vegetation or trees.

6.3.3   Temperature

It is well known that altitude can have an effect on temperature, generally causing a reduction of about 5 deg C per 1000 m (3 deg F per 1000 ft) but sometimes, due to local conditions, this gradient can be greatly magnified. As mentioned earlier (Section 6.2.1), gravity winds will bring cool air into a region and if this region is dammed in some way, by man or nature, the pool of air will continue to be cooled by radiation. These small areas are called frost hollows and the most famous of these is in the Austrian Alps some 100 km southwest of Vienna at an elevation of 1300 m. It is a depression of about 150 m in a plateau (Figure 6.5) and three successive winters recorded extreme

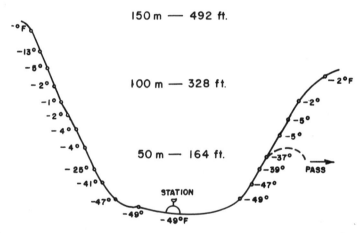

FIGURE 6.5   Temperatures on a chosen night in Gstettneralm frost hollow (Hawke, 1946)

minimum temperatures of −48, −51, and −51°C; at the same time the minimum temperature at the summit of Sonnblick, 3100 m, was −18°C. The conditions on the calm, clear and dry morning of 31 March, 1931, are given in Figure 6.5, showing differences of over 25°C in about 100 m. A similar, if not as dramatic, frost hollow is found near Rickmansworth in the Chilterns, England. This deep, sheltered valley, 55 m above sea level, has mean monthly minima very similar to those recorded in Braemar, Scotland, at 330 m elevation (Table 6.1). Freezes can occur in any month and on 7 July, 1941, the temperature ranged from 5·5°C (42°F) to 32·5°C (92°F).

TABLE 6.1

Mean monthly minimum temperatures (°C)

	J	F	M	A	M	J	J	A	S	O	N	D	Years
Rickmansworth	−2	−2	−2	1	4	6	8	8	6	3	0	−1	3
Braemar	−2	−2	−2	1	3	6	8	7	5	3	0	−1	2

Over bare soil, on a day with intense radiation, the air temperature will normally show a decrease with elevation, but on overcast days the air is nearly isothermal. The amplitude of the daily change is generally a function of the radiation and is therefore greater in summer than in winter, and on clear days as distinct from cloudy days. There is a time lag of maximum air temperature with elevation. At night if the sky is clear, the ground will cool more rapidly and the lower layers will become cooler, but about an hour after sunrise and near sunset isothermal conditions occur. When there is an increase of temperature with height, the condition is referred to as an inversion. The gradient of temperature over dry soil can be extreme, for with 70°C (158°F) on the soil there may be a reduction to 50°C (122°F) at a height of 1 cm or less (about 0·25–0·5 inches).

In a plant cover, the temperature patterns are determined greatly by the leaf orientation. In a cover of basically horizontal leaves, the uppermost tend to screen the lower, and an inversion results with the maximum temperatures occurring near the top leaves. For a vertical plant the radiation penetrates deeper, the effective radiative surface for both incoming and outgoing radiation occurring low down in the crop. Within a dense forest the pattern is somewhat similar to the horizontal leafed vegetation with maximum temperatures near the top of the canopy and sheltered conditions near the ground.

The nature of the surface is also extremely important as shown by a study in England, where radiation input is not very high. This indicated that for subsurface temperature in June, tar macadam could have a range of 32·6 deg C (42·6–10·0°C), sandy soil 26·0 deg C (35·1–9·1°C) and grass 16·0 deg C (29·3–13·3°C), while for air temperature the range was 14·2 deg C (21·8–7·6°C). It is clear that there will be a differential heating between urban and rural locations.

6.3.4 Soil Temperatures

On a clear day the radiation, and the temperature, at the soil surface reach a maximum at mid-day. Due to the physical characteristics of the soil, the heat takes a finite time to penetrate to a depth, z, and the maximum temperature at depth z, $_zT_{max}$, will be delayed. In addition, the heat becomes diffused and depleted so that the daily range at z, $_zT_R$, decreases with depth. The physical characteristic of greatest importance in this phenomenon is the thermal diffusity, K, of the soil, and Table 6.2 gives some characteristic values. Knowing the value of K,

TABLE 6.2

Thermal diffusivity of various materials

Material	Thermal diffusivity, $K (\times 10^3)$ cm² s⁻¹
Iron	260
Still air	150–250
Concrete	20
Rock	6–23
Wet clay	6–16
Wet sand	4–10
Dry sand	2– 5
Dry clay	0·5– 2

the approximate temperature pattern in the soil can be deduced from assumptions of the temperature wave at the surface. Some important points are given in Section 6·5.

6.3.5 Wind Speed Gradient

A study of many wind profiles from 10–200 cm (4 in–6 ft) over bare soil led Deacon (1949) to suggest that the rate of change of wind speed with height was related to $1/(height)^x$ where x varied with the thermal stratification of the air. When $x=1$, the atmosphere is in stable conditions, and the relationship can be used to obtain the mean wind speed at height z, $\bar{u}_z$

$$\bar{u}_z = c \, ln \, (z/z_0)$$

where c is a constant and z_0 is the roughness length. For example, z_0 is 0·001 cm for smooth mud flats, 9 cm for thick grass (50 cm tall) and 280 cm for a fir forest (550 cm tall). In high winds the vegetation tends to bend and this decreases the value of z_0.

A simple, empirical formula is

$$\bar{u}/u_1 = (z/z_1)^p$$

where p is about 0·15. However, p can vary widely with the roughness of the surface and the temperature gradient, from about 0·05–0·50.

6.3.6   Wind Flow

As has been noted earlier, the wind speed profile over bare soil is affected by the temperature stratification. While this is true above a plant cover or a forest, within the vegetation itself the air flow is reduced appreciably. The reduction occurring in a deciduous forest when the leaves come out is very noticeable. In forest clearings a reversal of air flow can often take place (Figure 6.6), a phenomenon dependent upon

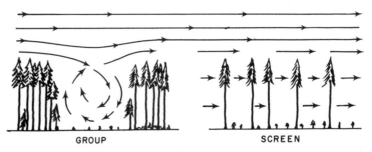

GROUP   SCREEN

FIGURE 6.6   Wind flow in wooded areas

the relationship of the clearing size to the tree height. Such clearings can often act like frost hollows, with cold air cascading from the tree canopy (*see* Section 6.3.3).

A drastic micro-scale wind phenomenon is the tornado, a feature almost exclusive to North America. Developed by the collision of two dissimilar air masses, warm moist air from the Gulf of Mexico and cooler dry air from the continent, these rapidly rotating whirlwinds approach with a noise like that of a train and their winds have been estimated at up to 1000 km h^{-1} (650 mph). At such speeds, straws can be driven through wooden planks and, in the path of a tornado, destruction is almost complete. Tornadoes are often spawned from hurricanes. Beulah, which hit the south Texas coast in 1967, gave rise to a record 115 'Twisters'.

6.3.7   Humidity

After sunrise, as evaporation from the soil increases and breaks the humidity inversion, absolute humidity will begin to rise. By late morning, when eddy diffusion increases, the water vapour content will decrease somewhat due to mixing, reaching a minimum in the afternoon. In the evening the humidity inversion is re-formed. In forested areas there is relatively little variation although a secondary maximum of absolute humidity is seen in the crowns, a situation that vanishes on overcast days when evaporation is generally less.

6.3.8 Rainfall

On the micro-scale, rainfall shows its greatest variations in forested areas where the leaf–bough configuration causes patterns of concentration and depletion. In a banana plantation during a convective rain of some 40 mm in two hours, the amounts recorded over an area 10 m (30 ft) square ranged from 7 mm to 93 mm. On a somewhat larger scale, a storm over the Burton Creek watershed in South Central Texas showed gradients of 50 mm (2 inches) in 1100 m (1200 yards) and 75 m (3 inches) in 1700 m (1900 yards).

6.4 Classification of Micro-climates

Recently, Gates (1973) has suggested a method of classifying micro-climates according to radiation, air temperature, wind and humidity. The number of subdivisions and their range, used for each of these four elements could be changed as required for a special study but, as Table 6.3 shows, this is a simple and effective technique for in Gates' proposed classification only 72 ($3 \times 4 \times 3 \times 2$) variations are possible. Such a method could be used for meso- or macro-climates to effect a broad zonation.

TABLE 6.3

A micro-climatic classification (after Gates, 1973)

Radiation (cal cm^{-2} min^{-1})		Air temperature (°C)		Wind (cm s^{-1})		Humidity (%)	
Sunny	1·0–1·6	Hot	30–50	Still	0–50	Dry	0–40
Cloudy	0·6–1·0	Warm	15–30	Breezy	50–200	Humid	40–100
Dark	0–0·6	Temperate	0–15	Windy	200 or more		
		Cold	below 0				

Situations	Conditions				Classification
Field at noon, clear, summer	Sunny,	hot,	still,	humid	SHSH
Field at night, cloudy, summer	Dark,	warm,	still,	humid	DWSH
Desert at noon, clear, summer	Sunny,	hot,	still,	dry	SHSD
Tree top at noon, clear, summer	Sunny	hot,	breezy,	humid	SHBH
Tree top at noon, clear, spring	Sunny,	warm,	breezy,	humid	SWBH
Tree top at night, clear, spring	Dark,	temperate,	still,	humid	DTSH
Tree top at night, cloudy, winter	Dark,	cold,	still,	humid	DCSH
Tree top at night, clear, winter	Dark,	cold,	windy,	dry	DCWD
Inside forest at noon, summer	Cloudy,	hot,	still,	humid	CHSH
Inside forest at noon, spring	Cloudy,	warm,	still,	humid	CWSH
Alpine tundra noon, clear, summer	Sunny,	temperate,	windy,	dry	STWD
Alpine tundra night, cloudy, summer	Dark,	cold,	breezy,	humid	DCBH
Lake shore noon, cloudy, summer	Cloudy,	warm,	breezy,	humid	CWBH
Room in house	Cloudy,	warm,	still,	dry	CWSD

6.5 Appendix

Assuming that the variation of the thermal properties of soil with depth is zero, the basic equations of heat flow in the soil are

$$\text{Heat flux, } Q = -k\frac{\delta T}{\delta z}$$

and

$$\frac{\delta Q}{\delta z} = -\frac{\delta(hT)}{\delta t}$$

where k is the thermal conductivity of the soil and h is the heat capacity. The rate of heat transfer depends upon k but the actual temperature change will be related to h.

From these equations

$$\frac{\delta T}{\delta t} = K\frac{\delta^2 T}{\delta z^2} \tag{1}$$

where

$$kh = K = \text{thermal diffusivity.}$$

Solving the above equation and assuming a sinusoidal temperature wave at the surface, $z = 0$, we find

$$R_z = R_0 \exp[-z(\pi/KP)^{1/2}]$$

where R_x is the temperature range at depth x, and P is the oscillation period (1 day or 1 year).

Also

$$(t_2 - t_1) = 0 \cdot 5 (z_2 - z_1)(P/K\pi)^{1/2}$$

where t_x is the time of maximum (or minimum) temperature at depth z_x. From (1) it is seen that the annual temperature range is reduced less rapidly with depth than the diurnal range, the ratio being $\sqrt{365}$, or about 19. It should be noted that K rises with increasing moisture content to a maximum, then decreases. Organic material lowers K but compaction increases K. Because of lower k and h, dry sandy soils heat more rapidly than dry clay soils and, conversely, cool more rapidly, especially in the surface layers. From the above equations specific ranges of temperatures and delay times can be calculated (Table 6.4).

TABLE 6.4

| Depth at which range is 1% of that on surface | | | Delay in thermal maximum | | | |
| | | | Daily | | Annual | |
$K\,(\text{cm}^2\text{s}^{-1})$	Daily	Annual	3 h	12 h	2 month	6 month
0·001	24 cm	4·6 m	4 cm	8 cm	1·0 m	3·0 m
0·008	68	13	12	24	2·9	8·7
0·012	84	16	14	28	3·6	10·8
0·025	120	23	20	40	5·1	15·3

References

Deacon, E. L. (1949) 'Vertical diffusion in the lowest layers of the atmosphere', *Q. J. R. Meteorol. Soc.*, **75,** 89–103

Gates, D. M. (1973) *Man and His Environment: Climate*, Harper and Row, New York, 175 pp.

Geiger, R. (1965) *The Climate Near the Ground*, Harvard University Press, Cambridge, Mass., 611 pp.

Hawke, E. L. (1946) 'Frost hollows', *Weather*, **1**(2), 41–5

Hudson, H. E., Jr., Stout, G. E. and Huff, F. A. (1952) *'Studies of Thunderstorm Rainfall with Dense Rainage Networks and Radar*, Illinois State Water Surv., Rept. Invest., 13, 30 pp.

Lott, G. A. (1953) 'The unparalleled Thrall, Texas, rainstorm', *Mon. Weather Rev.*, **81,** 195–203

Manley, G. (1946) 'Variations in the length of the frost free season', *Weather*, **1**(2), 60–1

7

Human Biometeorology

7.1 Introduction

Man has always been subject to the rule of the weather elements. In the earliest days, the human settlements began in regions where the climate was not too extreme; as fire, clothing and crude shelters were used, the limits of tolerance were extended. Nowadays, with controlled heating and air conditioning, a small volume of the atmosphere can be made habitable, if not pleasant, almost anywhere on the globe. However, in the face of such phenomena as tornadoes, hurricanes and floods, Man is still virtually powerless.

About fifty years ago a number of authors propounded such themes as *Civilization and Climate* (Huntington, 1924) and *Climate and the Energy of Nations* (Markham, 1947). Many of these theses on climatic determinism were conjectural and subjective, but we do know that through direct and indirect impact climate and weather do affect us and our way of living. Man is a very complex animal, and, because of the need to have a precise regulation of his body temperature, he is an organism in which it is very difficult to isolate cause and effect, due to the many inter-relationships, feedbacks and safety devices. In Table 7.1 a summary of human responses to thermal stress indicates some of the mechanisms introduced. The micro-climate inside Man, and the other high organisms, must be within certain narrow limits if the organs are to function efficiently and allow the animal to survive. Tolerance within is much less than the tolerance outside the body.

Slightly outside of the climatological realm is Man's inability, due to the lack of oxygen, to adapt to very high altitudes. It is well established that, even in the case of acclimatised persons such as the inhabitants of the Andean altiplano, permanent domicile above about 5200 m (17 200 ft) is not possible, although a religious settlement has been discovered at nearly 6300 m (21 000 ft). The highest city is Wenchuan in China at 5100 m (16 880 ft) and the highest observatory is Chacaltaya, 5400 m (17 800 ft) in Bolivia.

TABLE 7.1

Summary of human responses to thermal stress (Lee, 1958)

To cold	To heat
Thermoregulatory responses	
Constriction of skin blood vessels	Dilation of skin blood vessels
Concentration of blood	Dilution of blood
Flexion to reduced exposed body surface	Extension to increase exposed body surface
Increased muscle tone	Decreased muscle tone
Shivering	Sweating
Inclination to increased activity	Inclination to reduced activity
Consequential disturbances	
Increased urine volume	Decreased urine volume. Thirst and dehydration
Danger of inadequate blood supply to skin of fingers, toes, and exposed parts leading to frostbite	Difficulty in maintaining blood supply to brain leading to dizziness, nausea, and heat exhaustion.
	Difficulty in maintaining chloride balance, leading to heat cramps
Increased hunger	Decreased appetite
Failure of regulation	
Falling body temperature	Rising body temperature
Drowsiness	Heat regulating centre impaired
Cessation of heartbeat and respiration	Failure of nervous regulation terminating in cessation of breathing

7.2   Direct Effects

Solar radiation, especially in the ultraviolet range ($< 0.3\mu m$), causes tanning of the skin, blistering and, in certain prolonged exposure cases, skin cancer. In addition, the intense rays can lead to solar conjunctivitis and cataracts. On the benefit side, these rays develop anti-rickettsial compounds and devitalise some bacteria and germs. Radiation in the 0.32–0.65 μm range can also cause sunburn, so that even on overcast days, or behind glass, painful conditions can result.

Man is extremely tolerant to temperature alone; it is when a temperature extreme is in combination with other elements, such as wind and humidity, that it has the worst effects. However, under extremely cold conditions frostbite of the extremities can occur, while the cooling of the lungs at the intake of icy air can lead to a strain on the heart. Extremely high temperatures can lead to heat stroke and desiccation and, in conjunction with high humidity, can induce heat prostration and prickly heat, an unpleasant rash.

After all these 'negatives' the most obvious 'positive' is that mankind as a whole does continue to live and thrive!

7.3   Body Heat Balance

Because Man needs to maintain a reasonably constant deep-body temperature, under steady state conditions, heat gain must equal heat loss. Thus we may write

metabolic heat (M) + radiation (R_+) + convection (C_+) + conduction (P_+)
= radiation (R_-) + convection (C_-) + conduction (P_-) + evaporation (E)

When the air temperature is about 10°C (50°F) the radiative and convective loss is about 9 times that of the evaporative loss, but at 21°C (70°F) this is reduced to four times and, above 30°C (86°F), evaporation exceeds the radiation and convection loss. The average body temperature is 37–38°C (98–99°F), with 36–39°C (97–103°F) for efficiency, 32–41°C (90–106°F) for consciousness and 16–44°C (61–111°F) as extremes. The energy exchange pattern for man can be extremely complex under most environmental conditions—Figure 7.1 shows the many aspects playing a role in the problem. The problem is further compounded by the wearing of clothing, a condition that makes for varying absorptions, changes in resistance to air flow, evaporation and body heat loss.

Clearly, skin temperatures play a role in many of these components, and it is a complicating factor that, depending upon its location, our skin has a variable temperature. For instance, one study in Japan in winter gave toe and finger skin temperature at 15°C (59°F), the forearm at 25°C (77°F) and the waist at 34°C (94°F), but in summer the differences were reduced to about 2–3 deg C (4–5 deg F). Other investigators have related the average skin temperature (T_s) to dry bulb and wet bulb temperature (Figure 7.2).

7.4   Heat Gain

7.4.1   *Metabolic (M)*

Under normal resting conditions the average person generates about 50 kcal m^{-2} h^{-1}, a unit often called a Met ($= 0 \cdot 08$ ly min^{-1}). Light activity raises this to $1 \cdot 2$–2 Mets, moderate work to 3–5 Mets, and heavy work to 5–7 Mets (Table 7.2). The basal metabolic rate of $0 \cdot 8$–$1 \cdot 0$ Met is sufficient to raise the body temperature by 1 deg C (2

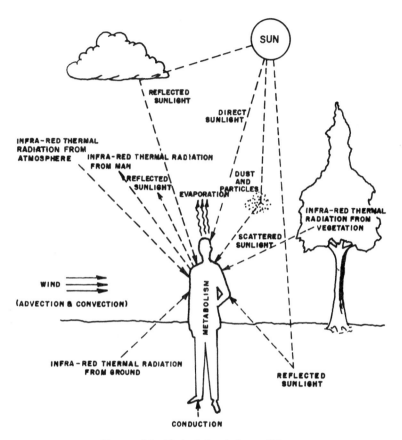

FIGURE 7.1   The body heat balance of Man

deg F) per hour if heat is not dissipated in some manner. Food is the main source of this energy, of which 80% goes in growth, body repair and heat production, with only 20% remaining for daily activities. The evaporative heat loss from the lungs due to breathing is about 25% so that the heat gain is actually only 0·75 times the metabolic heat.

7.4.2   Radiation (R_+)

The main source of gain is the sun and we must be careful to distinguish between the direct, unidirectional, beam and the diffuse, omnidirectional, input. On a clear day, the ratio (direct)/(direct plus diffuse) is related to the altitude of the sun, α, such that for $\alpha = 10°$ the ratio is 0·6, increasing to about 0·8 when α exceeds 45–50°. Figure 7.3

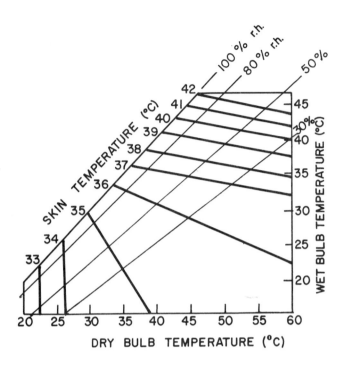

FIGURE 7.2   Skin temperature as a function of temperature and humidity. (Reprinted with permission from *Man, Climate and Architecture* by B. Givoni, Elsevier, 1969)

TABLE 7.2

Metabolic heat production related to human activities (Landsberg, 1969)

Kind of activity	No. of Mets
Sleeping	0·8
Awake, resting	1·0
Standing	1·5
Working at desk, driving	1·6
Standing, light work	2·0
Level walking, 4 km h⁻¹, moderate work	3·0
Level walking, 5·5 km h⁻¹, moderately hard work	4·0
Level walking, 5·5 km h⁻ ׳ with 20 kg load, sustained hard work	6·0
Short spurts of heavy activity (e.g. climbing or sports)	10·0

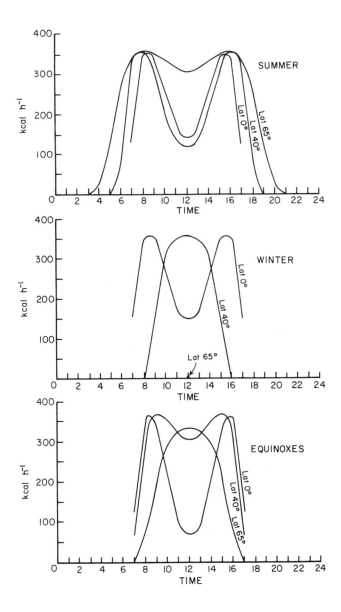

FIGURE 7.3 Radiation load on Man in latitudes 0°, 40°, 65°, at solstices and equinoxes (after Terjung and Louie, 1971)

shows how the direct radiation falling on Man varies through the day, at 0°, 40°N, and 65°N.

7.4.3   Convection (C_+)

A convective gain occurs only when the temperature of the advected air, T_a, exceeds the skin temperature, T_s. The gain has been shown to be related to $(T_a - T_s)$ and $(V)^{0.3}$, where V is the wind speed.

7.4.4   Conduction (P_+)

For a conductive increase to take place, part of the body must be in physical contact with a warmer surface. Generally, this is a minor part of the whole body heat gain but it may be of great physiological or psychological effect—for example, if one has a finger in a fire or is walking bare-footed on a very hot surface.

7.5   Heat Loss

7.5.1   Radiation (R_-)

Radiative heat loss occurs as a long wave interchange and is proportional to $(T_{skin}^4 - T_{surrounds}^4)$. At night, with a cold sky, this loss can be appreciable, and in hot areas this reversal of heat flow from the daytime gain can be a welcome relief during the darkness hours.

7.5.2   Convection (C_-)

In this case, following the same pattern as for the heat gain, the advected air must be at a temperature of less than T_{skin}.

7.5.3   Conduction (P_-).

The comments made for the conduction gain (P_+) are applicable to those for the conduction loss (P_-).

7.5.4   Evaporation (E)

Evaporation occurs from two main sources—the upper respiratory tract and the skin. It has been suggested that $E = aW(e_s - e_a)$ where a = water vapour transfer coefficient, W = wetted fraction of the body (effectively always greater than 0.1), e_s = saturation vapour pressure at the skin temperature, e_a = the vapour pressure of the air. It should be noted that when 1 gram of water is evaporated completely 0.58 kcal of heat are used.

Heat loss by evaporation through breathing is about four times the heat loss due to heating the air to the respired temperature. Man has approximately two million sweat glands and can sustain a loss of 2 litres h^{-1}

($= 14$ Met), but over 24 hours this is reduced to 0.5 litres h^{-1}. The ability to sweat increases during the hot months due to a physiological acclimatisation process. Immigrants to tropical areas generally begin to sweat before the long-term inhabitants. Natives of the tropics usually have a larger number of 'active' sweat glands than people from temperate regions.

Empirical relationships of the form $E_{max} - cV^x(42 - e_a)$, where c is a constant, V is wind speed and e_a = vapour pressure of the air, have been suggested, with x ranging from 0.3 to 0.6.

7.6 Heat Flow Equations

The basic equation of heat flow is

$$H_{xy} = (T_x - T_y)/I_{xy}$$

where H_{xy} is the heat flow between surfaces at temperatures T_x and T_y with a medium between the surfaces having an insulation of I_{xy}.

For the three 'layers' of interest here—body to skin, skin to clothing, clothing to air—the equation becomes

(a) $H_1 = (T_{body} - T_{skin})/I_{tissues}$
(b) $H_2 = (T_{skin} - T_{clothing})/I_{clothing}$
(c) $H_3 = (T_{clothing} - T_{air})/I_{air}$

An insulation unit often used is the CLO, which equals 0.18 deg C/kcal m^2 h or the approximate insulation of a business suit. It is found that the difference between body and skin temperatures should be greater than 2 deg C for efficient cooling and sufficient flow of heat to occur.

When no sweating takes place, it has been shown that, approximately,

$$H_1 = 1.21H_2, \quad H_2 = H_3$$

Also,

$I_{tissues}$ = 0.15 CLO (vasodilatation) to 1.81 CLO (vasoconstriction),
$I_{clothing}$ = 4.7 CLO in^{-1} (maximum), average about one CLO
I_{air} = 1.0 CLO (calm to 10 cm s^{-1} (0.2 mph) wind), reducing to 0.1 CLO at 25 m s^{-1} (50 mph).

If (b) and (c) are combined then

$$H = (T_{skin} - T_{air})/(I_{clothing} + I_{air})$$

For the nude body,

$$I_{cl} = 0, \text{ thus } T_a = T_s - H I_a$$
$$= T_s - \text{function (work, } I_a) \tag{1}$$

At low M, ($= 50$), $H = 38$ and the following examples show the effect of wind speed:

(a) with calm conditions, $I_a = 0 \cdot 18$, then $T_a = 33 - 38 \times 0 \cdot 18$
$$= 26°C (79°F)$$

(b) with a wind of $0 \cdot 2$ m s^{-1} ($0 \cdot 4$ mph), $I_a = 0 \cdot 14$, and
$$T_a = 33 - 38 \times 0 \cdot 14$$
$$= 27 \cdot 5°C (83°F)$$

(c) with a wind of $0 \cdot 5$ m s^{-1} (1 mph), $I_a = 0 \cdot 09$, and
$$T_a = 33 - 38 \times 0 \cdot 09$$
$$= 29 \cdot 5°C (85°F)$$

These are the values of T_a for which T_{body} remains constant. In other words, the naked man can maintain a constant body temperature at an environmental temperature of about 28–30°C.

When $H = 120$ (light work), with $0 \cdot 5$ m s^{-1} (1 mph) wind

$$T_a = 33 - 120 \times 0 \cdot 09 = 22°C (72°F)$$

so that a working, unclothed person can be in thermal equilibrium at only 22°C (72°F). For the clothed body, Equation (1) becomes

$$T_a = T_s - H(I_a + I_{cl})$$

If now the I_{cl} corresponding to a lightweight suit ($I_{cl} = 0 \cdot 18$ CLO) is used, with a low metabolic rate ($H = 38$) and a wind speed of $0 \cdot 5$ m s^{-1} (1 mph) ($I_a = 0 \cdot 09$), then

$$T_a = 33 - 38 \times 0 \cdot 27 = 23°C (73°F).$$

This example illustrates that, for these conditions, the lightly clothed man can rest comfortably at 23°C (73°F) whereas an unclothed person would have to be doing light work to remain in thermal equilibrium.

Clothing makes for increased complication of the equations, for not only do the colour and texture affect the amount of absorbed radiation, but the reaction of the material to moisture absorption and compaction can influence the insulation drastically. Cotton, when moist, becomes longer and easily compacted, whereas wool will continue to serve as a good air trap.

7.7 Climatic Indices

Temperature is not a perfect element for describing the integrated impact of the atmospheric environment upon man—although it may be the best single element. We have all experienced discomfort, due not so much to the air temperature alone, as to this in conjunction with the radiation, humidity or wind speed. Radiation generally acts to make for an apparent increase in the 'environmental' temperature; humidity's effect is greatest at high temperatures, whilst wind will lead to a sensation of a decreased 'environmental' temperature (except when the air temperature is greater than the skin temperature, i.e. about 32°C).

There are two special climatic regimes that have received the most attention in the study of the impact of the atmospheric environment on Man—the cold stress region and the heat stress region. In the cold stress region, shivering can increase the heat production to about 3 Mets but it is an exhausting activity. In addition, the movement increases convective heat loss. Vasoconstriction, which reduces skin blood flow, assists in controlling heat loss, but it increases blood pressure. Recent studies have shown that the Australian aborigines can sleep naked at nearly freezing air temperatures because they tolerate a falling body temperature, especially in their limbs. The Tierra del Fuegan Indians, living in an overcast, windy, very cold and wet environment, have a high heat production during the night—an ideal adaptation.

7.7.1 Cold Stress Region

The simplest approach used is one combining the wind speed and air temperature through the windchill factor. The heat loss (windchill index), K, is given by

$$K_{v,a} = (10\sqrt{v} + 10 \cdot 5 - v)(33 - T_a). \qquad T \text{ in } °C, v \text{ in m s}^{-1}$$

From this the windchill equivalent temperature, T_{we}, is calculated such that $K(v, T_a)$ is equal to $K(2 \cdot 2\, T_{we})$. In other words the cooling power of the atmosphere will be compared with that occurring at a wind speed of $2 \cdot 2$ m s^{-1} (5 mph). However, this approach, derived by Siple and Passel (1945), actually is applicable only to exposed flesh, and to improve it Steadman (1971) has devised a method that applies to clothed persons. The new windchill equivalent temperatures are shown in Figure 7.4, together with some for exposed flesh. As an example, at a temperature of $-10°C$ (14°F) and a wind speed of 5 m s^{-1} (11 mph) the Siple windchill equivalent temperature would be about $-19°C$ ($-2°F$) while that of Steadman would be $-13°C$ (9°F). Steadman's approach

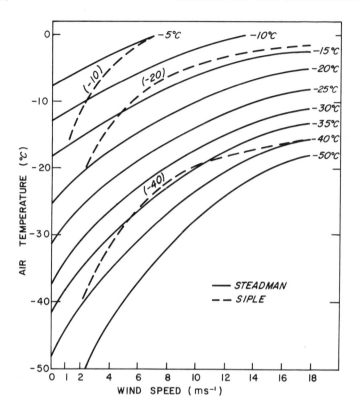

FIGURE 7.4   Windchill nomogram using Siple and Steadman equations

assumes that the person is clothed adequately to maintain thermal equilibrium.

7.7.2   Heat Stress Region

In this region, with temperatures in excess of about 25°C, attention must be focused on radiation and humidity, as well as temperature. Radiation poses a particularly intricate problem and, for a sophisticated energy budget approach, the reader is referred to Lowry (1969).

However, use is often made of the effective temperature, T_e, concept. This relates temperature–humidity combinations to their effect upon humans (Figure 7.5). For example, a temperature of 32°C (90°F) and a relative humidity of 20% would have the same value of T_e, namely 25°C (77°F), as a temperature of 26°C (79°F) and a relative humidity of 90%.

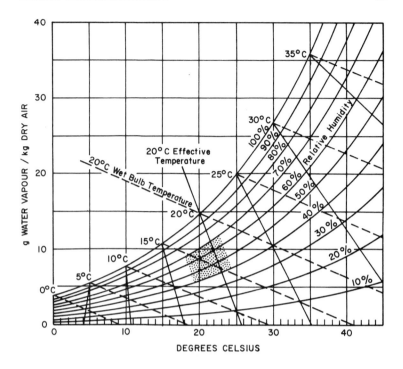

FIGURE 7.5 Psychrometric diagram with effective temperature. (Reprinted by permission of the World Meteorological Organization from W.M.O. 331, 1972)

An approximation to the formula is called the temperature–humidity index, THI, and is given by

$$\text{THI} = T_{\text{air}} - 0.55\left(1 - \frac{\text{r.h.}}{100}\right)(T_{\text{air}} - 58). \quad T \text{ in } °F$$

Maps of the effective temperature over the world in January and July have been given by Landsberg (1972). These show that values of about 25°C (77°F) exist around the equatorial belt, while readings below −20°C (−4°F) are found in central Siberia and Canada in January. The same publication gives a correction to be added to the effective temperature due to global radiation—it is approximately 1 deg C for each 0·1 ly min⁻¹. The corrections for global radiation and wind speed are given in Table 7.3.

TABLE 7.3

Wind speed (m s^{-1}) corresponding to a given global radiation and correction to
effective temperature

Global radiation	Correction (°C)							
	1	2	3	4	5	6	10	15
0·5 ly	16	4	2	1	0·7	0·5	—	—
1·0	—	16	7	4	2·5	1·2	0·7	0·5
1·5	—	—	16	9	6	2·5	1·5	0·7

7.8 Clothing Zones

Siple (in Newburgh, 1949) suggested a useful seven-zone division of the world using the concept of clothing requirements within each one. Naturally, the metabolic rate (work activity) will influence such a classification, as would the age, sex and physical condition of the person involved, but the concept is worthy of some consideration.

The first zone is that of minimal clothing (humid tropical), where the mean monthly temperatures vary between 20°C (68°F) and 30°C (86°F). The dictates here are traditional modesty, fashion, and protection. Normally a material such as light cotton will suffice. The second zone is the hot, desert type where high temperatures and radiation make necessary a clothing that protects from the solar rays, allows evaporative cooling and insulates against the generally appreciable night time cooling. Long flowing robes are the advised wear. The one-layer clothing zone (subtropical or optimum comfort area) exists where mean monthly temperatures are between about 10°C (50°F) and 20°C (68°F). Here, the arms and legs do not suffer unduly but the torso needs protection. Wool is ideal, with light cotton undergarments. In the rainy areas of this zone, as in zone one, an umbrella is a better protection than a waterproof material over the body, for the latter reduces both convective and evaporative loss.

The two-layer zone is in temperate, cool winter regions with mean monthly air temperatures between 0°C (32°F) and 10°C (50°F). Here, conditions are often humid and radiation is not excessive. The ideal clothing should allow about 6 mm (0·25 inch) of air to be trapped between the two layers, with the provision that the outer garment can be readily removed when the metabolic rate is increased (moderate to heavy work). The three-layer zone obtains where temperatures are between −10°C (14°F) and 0°C (32°F) and the four-layer between −20°C (−4°F) and −10°C (14°F). In the last case the Eskimo fur garments are perhaps the most efficient. Siple's last zone, the Arctic winter type, represents a region in which a comfortable heat balance cannot be maintained by clothing insulation alone.

Another classification has been developed for the needs of the US Army in the issuance of military clothing (Anstey, 1966). This also uses the temperature thresholds suggested by Siple.

7.9 Health

Man's health is affected throughout his life, in some measure, by the climate. In some of the more rigorous climates, the inhabitants develop an amazing tolerance to the average conditions, being able to walk barefooted on sand at temperatures of 75°C (about 170°F) or on ice, to live almost naked in near-freezing temperatures, or to bear intense heat loads without collapse.

Many of Man's illnesses are transmitted by certain vectors, vectors for which the weather conditions must be right to allow their survival or even, occasionally, become benign and bring about epidemic outbreaks. An obvious example of this is in the case of malaria, where the insect vector needs warm temperatures (22–25°C) and much rain (1000 mm per year) for water puddles in which to lay its eggs. Other diseases, such as sleeping sickness (transmitted by tsetse fly), yaws and bilharzia, etc., are climate-dependent, via the vectors.

Another realm of relationship between climate, weather and health concerns the direct effects on man. Landsberg (1969) has remarked

'Weather is proverbially fickle and each change in the weather requires an adaptation of the human body. Much of that occurs quite unnoticed by us. The physiological mechanisms are very efficient—up to a point. Yet, sometimes they break down and then trouble starts; a pathological reaction takes place that may result in injury, illness, or even death. The weather also affects our moods and steers some of our psychological responses.'

Ailments such as some forms of rheumatism and rheumatoid arthritis are apparently aggravated by cool, humid weather. Allergies to windborne material must be dependent on the weather conditions, as are pulmonary complaints, such as asthma where a rapid fall in temperature can increase the incidence of attacks. The variation in the incidence of heart infarctions with weather has also been studied, and the suggestion made that disturbed atmospheric situations will lead to a higher frequency of attacks than occur during spells of calm weather. Naturally, stresses due to unusual heat or cold will place a greater

burden on the circulatory system and be a potential source of trouble
for a heart patient.

7.10   Appendix

1. Steadman radiation increment is ΔT_R (°C), where

$$\Delta T_R = \alpha_s pG/h + b + (hbd_F/k_F)$$

where

α = weighted absorptivity of skin and clothing
p = proportion of body's surface effectively receiving radiation
   normally
G = direct insolation (cal m^{-2} s^{-1})
h = heat transfer coefficient (cal m^{-2} s^{-1} per deg C)
b = heat transfer coefficient referring to exhaled air (cal m^{-2} s^{-1}
   per deg C) = 0·18
d_F = fabric thickness (cm)
k_F = thermal conductivity of clothing (cal m^{-2} s^{-1} per deg C cm^{-1})

NOTE:

$\alpha \approx 0·8$ for skin; this varies with clothing but 0·8 is a reasonable
   approximation
p — shadow area proportion is given approximately by
   $4 \exp(-i/2)$ where i = solar altitude
$$h \approx 1·72 v^{0.75} + 0·0135 \left[4\left(\frac{T_A + 273}{100}\right)^3 + 0·3 \left(\frac{T_A + 273}{100}\right)^2 \right],$$
   where v is wind speed in m s^{-1}; T_A is air temperature in °C.
$d_F \approx$ depends on conditions
$k_F \approx 1·0$

Example
(a) Cold conditions, $p = 1/3$ (sun at 45°), $G = 160$, $h = 4$, $d_F \approx 1$,
   $k_F \approx 1·0$
   $\Delta T_R \approx (0·8 \times 1/3 \times 160)/[4 + 0·18 + (4 \times 0·18 \times 1)] \approx 10°C$
(b) Warm conditions, $p = 0·2$ (sun at 57°), $G = 240$, $h = 3$, $d_F = 0·1$,
   $k_F = 1·0$
   $\Delta T_R = (0·8 \times 0·2 \times 240)/[3 + 0·18 + (3 \times 0·18 \times 0·1)] = 12°C$

2. *Energy Budget*
Lowry (1969) shows how the energy budget type 4 can be rewritten
as

$$(T_c - T_a) = (M_B + M_{Hp})RA_S^{-1} + (a_s S + E)(h_c + h_r)^{-1}$$
$$+ (M_B + M_{Hp})A_s^{-1}(h_c + h_r)^{-1}$$

where $(T_c - T_a)$ is the difference between core or body temperature (37°C) and the air temperature; $(M_B + M_{Hp})$ is the metabolic output; R is the 'resistance' to conduction by the body and clothing; A_s is the total skin area in contact with the inner surface of the clothing; a_s is the absorptivity of the clothing to short wave radiation S; E is the latent heat transfer; $(h_c + h_r)$ is the sum of the convective and radiative transfer coefficients.

Example

Assumptions:

man—cylinder of diameter (body thickness), $D = 20$ cm; $L_B = 10$ cm; $a_s = 0.7$; $M_B/A_s = 0.083$ cal min^{-1}; $M_{Hp} = 0$; $E = -0.033$ cal cm^{-2} min^{-1} (68 cc h^{-1}); body conductivity $= k_b = 0.35$ cal cm^{-1} min^{-1} deg^{-1}

clothing—wool suit, $k_c = 0.005$ cal cm^{-1} min^{-1} deg^{-1}; thickness, $L_c = 1$ cm

environment—direct solar radiation, $S_d = 1.0$ cal cm^{-2} min^{-1}; $h_r = 0.86 \times 10^{-2}$ cal cm^{-2} min^{-1} deg^{-1}; altitude of sun, $\alpha = 30°$; wind speed, $v = 10^4$ cm min^{-1} (nearly 4 mph)

Intermediate steps: $R = L_c k_c^{-1} + L_b k_b^{-1} = 10/0.35 + 1/0.005 = 228$;
$h_c = (2 \cdot 2 \times 10^{-3})(V)^{1/3}(D)^{-1/2} = 10^{-2}$;
$(h_c + h_r)^{-1} = 1 \cdot 86 \times 10^{-2} = 54$;
$S = S_d \cos \alpha/\pi = 0.25$

Calculations:

$(M_B + M_{Hp})RA_s^{-1} = 0.083 \times 228 = 19.0°C$ (1) metabolic-clothing component

$a_s S(h_c + h_r)^{-1} = 0.7 \times 0.28 \times 54 = 10.4°C$ (2) radiation component

$E(h_c + h_r)^{-1} = -0.033 \times 54 = -1.8°C$ (3) evaporative loss component

$(M_B/A_s)(h_c + h_r)^{-1} = 0.083 \times 54 = 4.9°C$ (4) metabolic component

Thus, $T_c - T_a$ $= 32.5°C$

Therefore, with a $T_c = 36°C$ the interpretation is that this man in a wool suit would be comfortable (in thermal equilibrium) at $3 \cdot 5°C$ (38°F). Without clothing the component (1) becomes $(0.083)(10/0.035) = 2.3$ and $T_c - T_a$ is reduced to $15.8°C$; that is, thermal equilibrium at about 20°C (68°F). If, in addition, the sun were not shining, $T_c - T_a$ becomes $5.4°C$, to give thermal neutrality at 30°C.

References

Anstey, R. L. (1966) *Clothing Almanac for Southeast Asia*, Tech. Report 66–20–ES U.S. Army Natick Labs., 35 pp.

Givoni, B. (1969) *Man, Climate and Architecture*, Elsevier, New York, (Architectural Science series), 364 pp.

Gregorczuk, M. and Cena, K. (1967) 'Distribution of effective temperature over the surface of the Earth', *Int. J. Biometeorol.* 11 (2), 145–50

Huntington, E. (1924) *Civilization and Climate*, Yale University Press

Landsberg, H. E. (1969) *Weather and Health*, Doubleday, 148 pp.

Landsberg, H. E. (1973) *The Assessment of Human Bioclimate, a Limited Review of Physical Parameters*, W.M.O 331, Tech. Note 123, 36 pp.

Lee, D. H. K. (1958) 'Proprioclimates of man and domestic animals in climatology; reviews of research, U.N.E.S.C.O.' *Arid Zone Res.*, **10**, 102–25

Lowry, W. P. (1969) *Weather and Life*, Academic Press, New York, 305 pp.

Markham, S. F. (1947) *Climate and the Energy of Nations*, Oxford University Press, Oxford

Newburgh, L. M. (1949) (Ed.) *Physiology of Heat Regulation and Science of Clothing*, Saunders, Philadelphia

Siple, P. A. and Passel, C. F. (1945) 'Measurements of dry atmospheric cooling in subfreezing temperatures', *Proc. Am. Philos. Soc.*, **89**, 177–99

Steadman, R. G. (1971) 'Indices of windchill of clothed persons', *J. Appl. Meteorol.* **10**, 674–83

Terjung, W. H. and Louie, S. A. (1971) 'Potential solar radiation climates of Man.' *Annals Assoc. Am. Geogr.* **61**, 481–500

Suggested Reading

Edholm, O. G. (1966) 'Problems of acclimatization in man', *Weather*, XXI (10), 340–9

Licht, S. (1964) (Ed.) *Medical Climatology*, E. Licht, New Haven, Conn., 753 pp.

Macpherson, R. K. (1962) 'The assessment of the thermal environment', *Br. J. Ind. Med.*, **19**, 151–64

Tromp, S. W. (1963). (Ed.) *Medical Biometeorology*, Elsevier, New York

8

Climate and Building

8.1 Introduction

Man has shown how to live anywhere on earth if energy, of the right sort, can be made available to modify the immediate environment to bring it within his tolerance limits. However, the real need, and the challenging problem, is to design for, and not in spite of, the climate. A recent article (Stein, 1973) has drawn attention to the present waste of energy in buildings, and quotes

> Since Lever House was built on Manhattan's Park Avenue in 1952, glass-clad buildings have sprung up all across the nation. Most are energy hogs because glass is a notoriously poor insulator. Heat loss could have been cut by half had double-glazing been used (that is, two panes of glass hermetically sealed with an air space between them which acts as an insulator). Heat gain can be reduced by using the new reflective metallic glass, which substantially blocks solar heat and light. What is true for an office or apartment building is true for a home.

The article also notes

> Dublin, a New York engineer, has calculated that a building uses 29 per cent less energy for cooling if the broad sides face north and south. Why is it, asks Dublin, that all sides of a building are often treated as if they were the same? Why not have no windows on the west side, and fewer in the corners? His point is simply that energy use must be factored into building design, and that means starting at the beginning.

Another attempt to quantify the saving is given:

> [He] claims that an average office building is occupied 3100 hours annually with 500 hours in the temperature range where untreated outdoor air could be used. Simply opening the windows would bring about 19 per cent reduction in the

use of energy for handling air—but how many office buildings have windows that can be opened?

Man's ideas about housing have been rather unscientific for many years; consider the remark of Lawrence (of Arabia) to a friend who was going to the tropics,

> Do not fall into the Khartoum fault of wide streets. In the tropics, air (fresh or foul) is an energy; also sunlight. You want houses of immense height and vigorous overhang. Streets like alleys, half dark and full of turnings to exclude the winds.

As Lawrence's friend was to live in the highlands of Kenya the comment is very misleading. In addition, many ex-colonial territories generally have examples of houses modelled on the 'mother country' style—a decision that often leads to uncomfortable homes. It was not only the Europeans that fell into this trap, the Omani Arabs when they moved to hot, humid Zanzibar built palaces fitted for the hot, dry area they had left.

When a building is erected it alters the micro-climate of the area—not only of the area it covers physically, but also the habitat around it. This is especially true for air flow, as is seen in Figure 8.1 where the effect of different roof pitch is shown. The interesting aspect to note is that the building can induce a flow towards the leeside wall; this can help in ventilating outdoor and play areas.

High rise buildings are particular changers of air flow patterns and they can bring about windy conditions at their base. In areas of tall buildings (city centres) the channelling effect is often unpleasantly noticeable.

8.2   Aspects of Major Impact

In a manner similar to the studying of the effect of the atmosphere on Man (Chapter 7) four aspects of major importance in building problems can be identified: thermal; air flow; illumination; dampness.

8.2.1   Thermal
This aspect deals with both temperature and radiation and the major problem concerns heat transfer. The ultimate aim is to develop an ideal 'cryptoclimate', one with comfortable thermal range and maximum and minimum temperatures occurring at right times. This facet is considered on page 90. An investigation in which the air temperature reached 28°C

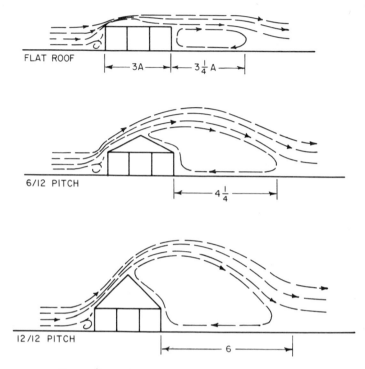

FLAT ROOF

|◄— 3A —►|◄— 3¼ A —►|

6/12 PITCH

|◄— 4¼ —►|

12/12 PITCH

|◄— 6 —►|

FIGURE 8.1   Air flow around a building (after Evans, 1957)

(81°F) showed that the east wall reached 61°C (142°F) at 9 a.m., the west wall reached 63°C (145°F) at 4 p.m., the south wall 48°C (120°F) at midday and the north wall 37°C (99°F) between 10 a.m. and 4 p.m. At about 6 a.m. and 8 p.m. the air and four walls were all at about the same temperature, 24°C (75°F) and 26°C (79°F) respectively. At night the walls were about 2 deg C cooler than the air temperature. The total radiation load on a house is generally not too dependent upon orientation, although the optimum (most heat in winter, least in summer) is found with the long walls running east–west. More important is the location of the rooms so that they receive (or do not receive) radiation at the right times of day. Most of the radiation load is received on the roof so that the colour of the roof is of major importance. A study showed that ceilings under a grey roof could be as much as 6°C (11°F) warmer than those under a whitewashed roof. Within the house the importance of keeping a cool ceiling (i.e. good roof insulation) and having high rooms is dramatically shown in Table 8.1.

TABLE 8.1

Radiation from ceiling at different temperatures, kcal h^{-1}
(Givoni, 1969)

Ceiling height (m)	Ceiling temperature (°C)			
	35·0	40·5	46·1	51·7
2·4	5·7	17·7	30·2	44·0
3·6	3·7	11·7	20·2	29·0
Difference due to 1·20 m reduction	2	6	10	15

Radiation penetration into a room can be estimated, approximately, using Table 8.2 which ilustrates the greater efficiency of the external shading devices over the internal types.

8.2.2   Air Flow (Pressure on Building and Ventilation)

The importance of wind effects on building is evidenced by the number of conferences and publications devoted to this subject in recent years. A few years ago the proceedings of a special meeting (National Research Council, Canada, 1968) ran to over 1200 pages

TABLE 8.2

Shading effectiveness

Coefficient = 100 − % transmission

Coefficient	
0	Clear window
9	Inside dark roller, half drawn
19	Inside dark roller, drawn; inside medium roller, half drawn
25	Inside dark blind, drawn
29	Inside light roller, half drawn
34	$\frac{1}{4}$″ heat absorbing glass; inside medium blind, drawn
38	Inside medium roller, drawn
42	Dark grey heavy drapes
45	Inside white blind, drawn
40–50	Tree performing light shade
53	Light grey heavy drapes
57	Outside awning, two-thirds drawn
60	Inside white roller, drawn; off-white heavy drapes
72	Outside aluminium shading screen
75	Outside canvas awning (dark or medium)
75–80	Dense tree performing heavy shade
85	Outside white awning, drawn

and a remark in the Foreword is worth noting: 'Much remains to be done in this field not only in theoretical and experimental research, but also in conveying to the members of the design professions in engineering and architecture the vital importance of understanding the action of wind on structures, especially in connection with taller buildings and slender structures.'

There are two distinct implications of wind effects, one due to the pressure of high winds on the structure, the other dealing with naturally induced ventilation within the building.

8.2.2.(a) Pressure   The pressure exerted by the wind on a surface is proportional to the square of the wind speed, thus at 15 m s⁻¹ (33 mph) the pressure is about one thousand times greater than at 0·5 m s⁻¹ (1 mph). Hence the devastating effect of hurricanes, typhoons and tornadoes with winds recorded in excess of 45 m s⁻¹ (100 mph). When the wind blows on the face of a building a greater pressure results, but in the lee a lower pressure occurs. In the cities the proximity of buildings to each other tends to reduce the 'standard' pattern of increase of wind speed with height (Chapter 6) and experiments suggest that the lower floors experience about one-third of the free air flow; this increases to two-thirds for the middle floors and suburban sites while at upper floors the factor becomes unity.

8.2.2.(b) Ventilation   Ventilation of a building is often effected by artificial means; it is used to remove odours and vitiated air, to give thermal comfort by increasing heat loss from the skin and to cool the structure of the building. When natural ventilation is used two items must be considered—the pressure pattern around the building and the inertia of the air. A prime factor in natural ventilation is the location of windows and doors. It has been shown that better ventilation results when the air stream has to change direction in the room than if the flow is directly across. For increased ventilation in a room two openings, an inlet and outlet, are necessary and greater average air flow is only achieved if both openings are increased in size. When partitions or walls are introduced, as does occur in most houses, the resultant flow becomes rather complex and the interested reader should refer to Givoni (1969). Some detail of the amount of ventilation needed is given in Section 8.6.

8.2.3 Illumination

This is very often related to radiation but the emphasis in this section is on the visible radiation. Often diffuse (sky) light is preferable to the direct (heating) radiation unless the latter occurs with the sun at low altitude, i.e. near sunrise or sunset. Skylights and clerestory windows can be of real benefit. Light surfaces (and mirrors) can be used at selected places indoors to increase reflection and make for a bright, yet cool, house. In trying to reduce light (and radiation) penetration, various shading devices can be used. The size of the overhang and fin width can be calculated (using solar position data, page 12) to give the shelter desired.

8.2.4 Dampness

The two divisions here are between rainfall and snowfall. Ideally, the overhangs of a house should shelter much of the wall area, while guttering should be of adequate size to cope with the average heavy storm of the region. However, in many parts of the world the concentration of water into a downspout only causes severe localised flooding or washaway, for storm sewers do not connect to house drainage. In this situation an argument can be made for guttering only the entry areas. Shrubs placed away from the walls (too close will lead to damp spots) will also shield the building from rainfall penetration. Snow build-up can cause an appreciable load to be placed on the roof trusses unless the roof pitch ensures minimal accumulation.

8.3 Indoor Climates

Climates of enclosed spaces are referred to as cryptoclimates and it is house cryptoclimates that are of special interest in this section. The existence of the house radically changes some elements and even excludes others, such as precipitation and wind (except for ventilation, *see* Section 8.2.2(b)). The elements of most interest within the house are temperature and humidity, and these are influenced by the external radiation, temperature and humidity.

The effect of orientation, thickness of wall and colours, on the internal surface temperature is shown in Table 8.3, while Figure 8.2

TABLE 8.3

Information on internal wall temperatures (°C)

Wall thickness	Colour	Range (°C)	Delay in time of maximum temperature
10 cm	Grey	9 (north) to 14 (west and south)	
	Whitewashed 5		about 4 hours
20 cm	Grey	4 (north) to 6 (others)	
	Whitewashed 2		about 8 hours

The daily air temperature range was 6°C.

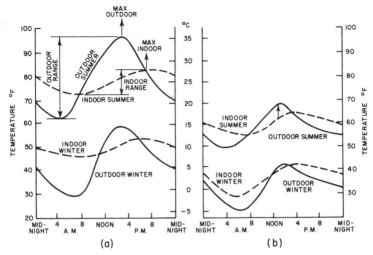

FIGURE 8.2   Comparison of typical internal and external temperature in (a) an adobe in hot, dry climates and (b) a wood structure in middle latitudes. (Reprinted by permission of John Wiley and Sons from *Climate and Man's Environment* by John E. Oliver, 1973)

gives the patterns obtained in two latitudes with two different wall materials. From physical considerations, it can be shown that when it is required to induce a time lag of 12 hours in the extremes of temperature (warmest then occurs indoors at about 3 a.m. and coolest at 6 p.m.) wall thickness should about 19 cm with insulating wood fibres, 22 cm with wood, 34 cm with brick, 39 cm with concrete. Such thicknesses will then reduce the inside temperature range to about one twenty-fifth of the outside range. Such large thicknesses are usually unrealistic, and even with enclosed air spaces the lag is reduced to only 3–6 hours. Of course, the outer surfaces can reach extreme temperatures. Landsberg cites that on one occasion with air temperature 25°C (77°F), wooden boards reached 58°C (136°F) and tar on waterproofing was at 88°C (190°F).

The general range of variation in the cryptoclimate as related to outdoor conditions is shown in Table 8.4. It must be noted that the given ranges are so wide because of the consideration which has to be given to the many different materials used in construction.

8.4   Native Housing

An interesting article on 'Primitive Architecture and Climate' (Fitch and Branch, 1960) made the very pertinent comment that '. . . the worst

TABLE 8.4

Range of variation in indoor climate (Givoni, 1969)

The climatic variable	Range of variation
Solar radiation absorbed in the walls	15–90% of incident radiation
Solar radiation penetrating through windows	10–90% of incident radiation
Indoor air temperature amplitude	10–150% of outdoor amplitude
Indoor maximum air temperature	−10 to −40°C from outdoor maximum
Indoor minimum air temperature	0 to −7°C from outdoor minimum
Indoor surface temperature	−8 to −30°C from outdoor maximum and minimum
Average internal air speed, windows open	15–60% of outdoor wind speed
Actual air speeds at any point in room	10–120% of outdoor wind speed
Indoor vapour pressure	0–7 mm Hg àbove outdoor level

he (the modern architect) faces is a dissatisfied client. When the primitive architect errs, he faces a harsh and unforgiving Nature.' In a variety of climatic zones around the world the local inhabitants have evolved a structure well adapted to the environment, often with characteristics we should do well to embody in our buildings, perhaps with improvements through the use of better technology. Basically there are seven regions in which native housing has incorporated the fundamental features for 'living with the climate'.

8.4.1 Hot, Humid Zone

In this area the only alleviation to the enervating conditions is to use to their maximum extent whatever breezes may occur. Houses are raised and are of only one room in depth. Shutters are used to protect against the seasons of intense rainfall and within the home removable or hanging screens are common (Figure 8.3). The raised house allows for some protection against insects as well as giving the living area a greater air flow. The area beneath the dwelling should be kept clear and not, as the author has seen all too frequently, boarded in for extra habitation or a garage, nor even is it to be used for a sheltered conservatory in which the mistress of the home can keep her numerous plants. In India a form of ventilation shaft (funnel) is used to induce an air flow into the house.

8.4.2 Hot, Dry Zone

The problem here is to protect from the intense radiation and, often, to keep out blowing sand. Dried mud bricks are frequently used as they are good insulators, and houses are built close together to shield out much sunshine. Small windows, white roofs and walls are common and a flat roof makes a very pleasant place for the evening meal (Figure 8.3).

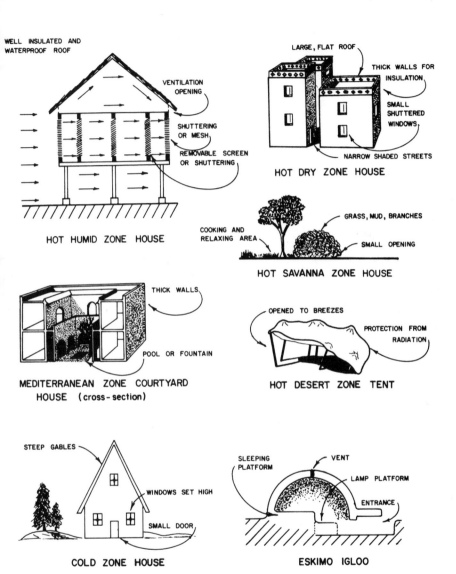

WELL INSULATED AND
WATERPROOF ROOF

VENTILATION
OPENING

SHUTTERING
OR MESH

REMOVABLE SCREEN
OR SHUTTERING

HOT HUMID ZONE HOUSE

LARGE, FLAT ROOF

THICK WALLS FOR
INSULATION

SMALL
SHUTTERED
WINDOWS

NARROW SHADED STREETS

HOT DRY ZONE HOUSE

COOKING AND
RELAXING AREA

GRASS, MUD, BRANCHES

SMALL OPENING

HOT SAVANNA ZONE HOUSE

THICK WALLS

POOL OR FOUNTAIN

MEDITERRANEAN ZONE COURTYARD
HOUSE (cross-section)

OPENED TO BREEZES

PROTECTION FROM
RADIATION

HOT DESERT ZONE TENT

STEEP GABLES

WINDOWS SET HIGH

SMALL DOOR

COLD ZONE HOUSE

SLEEPING
PLATFORM

VENT

LAMP PLATFORM

ENTRANCE

ESKIMO IGLOO

FIGURE 8.3   Native housing. (Reprinted with permission of Oxford University Press
from *Applied Climatology: An Introduction* by John F. Griffiths, 1966)

8.4.3 Hot, Savanna Zone

This zone combines the climatic characteristics of zones 8.4.1 and 8.4.2 and dwellings are generally constructed of mud and grass (Figure 8.3) in the shade of a spreading acacia tree. In some areas, skins are also used. In central Nigeria the dome is of two parts, the outer one of thatch supported by pegs placed in the inner hemisphere of mud. This type of construction, with its trapped air space, is a great improvement in the single model depicted. In central Tanzania the Irakwa have become troglodytic, living below ground in a cooler, moister atmosphere.

8.4.4 Mediterranean Zone

This region, with hot dry summers and warm wet winters, is generally regarded as one of the most beneficent for mankind. Older houses have courtyards with fountains or pools to modify the heat and low humidity of the summers; because the sun is never in the zenith there are few direct rays entering the courtyard (Figure 8.3). However, at night if the opening is large enough, it can become an artificial frost hollow.

8.4.5 Sub-tropical Deserts

The nomads of these areas, in which some of the highest surface temperatures are recorded, have little except animal skins as protection but even these can be used efficiently (Figure 8.3). The side flaps can be raised readily to channel any breezes, or lowered when sand storms begin. The adobe brick dwellings and the homes cut into the mesa walls, found in the south-west of the USA, are also ideal for this type of climate.

8.4.6 Cool Regions

In this zone, the radiation–temperature regime very rarely places a high heat load on the house, and the problem is really to conserve heat. Thus, homes are small and good insulation is essential. The sod huts and small houses of the European peasantry, whilst smoky and noxious, must have been more comfortable than the vast, rambling castles and fortresses of the wealthy.

8.4.7 Cold Regions

In the densely forested parts of this region wooden huts were, and still are, general. The roofs are steep to prevent an excessive accumulation of snow and ice (Figure 8.3). On the plains the Siberian Indians used a double tent, a yurt, which gave a reasonably comfortable inner sanctum, a luxury not developed by the Indians of North America who suffered in their single tepee during the long winters. In the frozen north the Eskimos developed a perfect solution, the igloo (Figure 8.3), constructed of snow blocks, or using these over a framework. The internal temperatures, raised by oil lamps and the Eskimos themselves, may run as much as 36°C (65°F) above external temperatures.

8.5 Landscaping

Not only should the garden or yard be developed to give pleasure to the dweller, but it should both assist the house to blend with its micro-environment and make the cryptoclimate more pleasant by natural means.

Shading of the roof and walls by deciduous trees, of the right height and spread, is often ideal, for in winter when the leaves have fallen, the less intense radiation can penetrate and is available for home heating and extra light. In very dry areas, plants can be instrumental in increasing the humidity while in wet areas certain flora (such as willow and fig trees) could be used to transpire excessive moisture from the soil to the air. Of course, too many trees and shrubs will only seek to induce air stagnation and can result in an unpleasant environment. Indeed, wind speed and direction are radically changed both by constructions and introduced vegetation, a fact that can be used to shelter areas from cold air flow and to channel cooling breezes during the summer. An outstanding paper on this problem of the effect of landscape development on natural ventilation is by White (1954) who used models in a special wind tunnel.

Grass or ground cover will keep the area around the house cooler than slabs of concrete, as well as assisting in making for more uniform drainage and soil moisture patterns. In addition, the glare or reflection will be less over vegetation than over cement. Of course, it must be recalled that night time conditions tend to be cooler and more humid over the vegetation covers.

8.6 Appendix

Ventilation Index, E (Webb, 1957)
It has been shown empirically that

$$E = 7.72 - 6 \log \delta T$$

where δT is the air temperature rise and

$$(\delta T)^3 = 0.045 \, (400 \, p + X)^2 / A^2 h,$$

where p is the number of persons present (assumed to be at rest), X is the aggregate rate of heat production of all other heat in the room (in B.t.u.),

A is the total area of the openings to the outside (ft^2), h is the lesser of the height of ceiling or top of the highest opening (ft).

Table 8.5 shows the qualitative meaning of E and Figure 8.4 depicts a graphical method of obtaining E, since

$$E = 4 \log A + 2 \log h - 4 \log p - 4 \log (1 + x/400\, p)$$

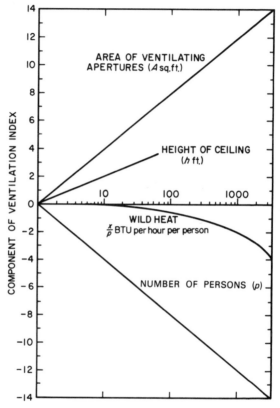

FIGURE 8.4   Graphical determination of ventilation index E (Webb, 1957)

References

Evans, B. H. (1975) 'Natural air flow around buildings', *Texas Engineering Expt. Stn., Res. Report*, **59**, 15 pp.

Fitch, J. M. and Branch, D. P. (1960) 'Primitive architecture and climate', *Sci. Am.*, **207**, 134–44

TABLE 8.5

Ventilation Index E	Rise of temperature ΔT deg F	deg C	Comment
10	0·4	0·2	Excellent, cool
8	0·9	0·5	Very good
6	1·9	1·1	Good
4	4·2	2·3	Adequate
2	8·9	5·0	Uncomfortable
0	19·3	10·7	Dangerous

Givoni, B. (1969) *Man, Climate and Architecture*, Elsevier, New York, 364 pp.

Nat. Res. Council, Canada (1968) 'Wind effects on buildings and structures', *Proc. Int. Res.* Seminar, University of Toronto Press, Vols. 1 and 2

Stein, J. (1973) 'There are ways to help buildings conserve energy', *Smithsonian*, **4,** 7

Webb, C. G. (1957) 'Natural ventilation in low-latitude buildings', *J. R. Inst. Br. Archit.*, **64,** 1–3

White, R. F. (1954) 'The effect of landscape development on the natural ventilation of buildings and their adjacent areas', *Texas Engin. Expt. Stn., Res. Report*, **45,** 16 pp.

Suggested Reading

Aronin, J. E. (1956) *Climate and Architecture*, Reinhold, New York

Conklin, G. (1958) *The Weather Conditioned House*, Reinhold, New York, 238 pp.

Griffiths, J. F. and Griffiths, M. J. (1969) *Bibliography of Weather and Architecture*, U.S. Dept. of Commerce, E.S.S.A. Tech. Mem.—EDSTM 9 72 pp.

Olgyay, V. (1967) *Design With Climate*, Princeton University Press, Princeton, Mass., 190 pp.

Olgyay, V. (1967) 'Bioclimatic orientation method for buildings', *Int. J. Biometeorol.*, **11**(2), 163–74

Page, J. K. (1964) 'Indoor climate in arid and humid zones. Summary report on Symposium on indoor climate in arid and humid zones', *Int. J. Biometeorol.*, **8**(2)

9

Climate and Agriculture

9.1 Introduction

When man began to develop a settled life, agriculture was made possible. Even in its primitive beginnings, the new 'farmer' would soon develop an awareness of seasonal rhythms, for on these the success or failure of his food supply depended. In the many thousands of years since then, we have still come only a small way in our understanding of the interrelationship between crops and climate (and weather). To a large extent, the farmer is still subject to the vagaries of the weather and, unfortunately, he can do little to nullify their damaging effects, a modification of the destruction being the best he can achieve.

The study of agrometeorology, which needs close cooperation between agricultural specialists and meteorologists, has many facets but most of the studies fall under one or more of eight categories:

1. Climatological—discovering analogous areas for crop introduction; calculating averages and probabilities; suggesting average planting dates, etc.
2. Meteorological—preparing and disseminating special forecasts for freezes, rainfall, high temperatures combined with high humidities, etc.
3. Meso-climatic—studies of local variations as these affect crops and animals.
4. Micro-climatic—investigations of actual atmospheric conditions within a crop.
5. Micro-habitat—usually of a special research nature concerning a single plant or even a single leaf.
6. Pests and diseases—the atmospheric influence on the vectors of plant and animal diseases.
7. Modification of the atmosphere—the measurement or estimation of the attempts to modify the environment to one more suitable for plant or animal (Chapter 10).
8. Analytical—the calculation of time of maturity and/or production of crops or animals, the detailed modelling of the plant within its environment.

As yet, there is a long way to go in all these divisions, but energy demands are putting pressures on (8) when there is insufficient understanding of the cause and effect relationship.

9.2   Climate and Soil

With the exception of hydroponics, crops are grown in soil and it is logical, therefore, to begin our study with a consideration of this medium. Soil is the end product of five factors—parent material (bedrock), climate, relief, plants and animals, and time. Soils are then considered as of three types:

1. zonal, where there has been appreciable climatic and biological influence;
2. intrazonal, where bedrock and relief dominate, and
3. azonal, where time has been too short or soil has been transported to the area.

Climate has a direct influence mainly through temperature and rainfall. These two elements affect the amount of humus, the decayed organic matter in the soil that is so essential in the forming of solutions to enable plants to utilise certain chemicals. Heavy rainfall can leach the soil, removing mineral or organic substances and, with high temperatures, can bring about laterisation, the rapid removal of silica. In semi-arid and arid regions the upward movement of soil moisture, due to high evaporation rates, can cause an undesirable concentration of salts in the upper layers.

Soils are also eroded by the effect of climatic elements. High intensity rainfall can lead to sheet or gully erosion whilst strong winds can remove dry topsoil, such as occurred in the Dust Bowl days of the 1930s in the USA, and as happens frequently in the deserts of the world. More erosion occurs as rainfall increases and is greatest in the temperate areas, for in many tropical areas the dense vegetation helps to bind the soil.

Temperature can affect relief by its diurnal cycles—weathering of rocks occurring due to the regular heating and cooling of the surface—or, in certain areas, a regular freeze/thaw pattern that results in particle sizing. The phenomenon of frost-heave, when water in the soil freezes, can lead to marshy conditions, while intense cold will lead to a permafrost condition. A continuous layer of permafrost generally exists when the mean annual temperature is below about $-5°C$ ($23°F$).

9.3   Climate and Vegetation

There is a definite pattern of correlation between natural vegetation and the elements of temperature and rainfall. It is not a perfect relationship, for soil variation and other factors are also important; nevertheless, the tie-in has led to many classifications aimed at a climate–vegetation link.

In the hot areas, adequate moisture will lead to a rain forest (often incorrectly called jungle) but if a short dry spell occurs, a deciduous forest, with a vegetative resting period, will result. As rainfall decreases, the forest will give way to a thorn woodland, savanna (tall grassland), scrubland, desert grasses and, finally, desert. In temperate zones, the wettest parts will support a deciduous forest, and as the dry season increases, sclerophyllous woodland (summer drought), prairie, steppe and desert occur, in that order. In cool zones, sufficient rainfall will give rise to the taiga or northern coniferous forest and, as temperatures decrease, the tundra region and perpetual snow and ice follow.

9.4   Plant Response to Radiation

9.4.1   General

In 1951 the Dutch Committee on Plant Irridation proposed that solar radiation could be considered under eight separate sections with respect to its impact on plant life. These wave bands are as follows:

1. greater than $1 \cdot 0$ μm—no effect
2. $1 \cdot 0 – 0 \cdot 72$ μm—causes elongation, photoperiodic responses
3. $0 \cdot 72 – 0 \cdot 61$ μm—chlorophyll absorption and photoperiodic responses
4. $0 \cdot 61 – 0 \cdot 51$ μm—no significant response
5. $0 \cdot 51 – 0 \cdot 40$ μm— strong chlorophyll absorption band
6. $0 \cdot 40 – 0 \cdot 31$ μm—shortens the plant and thickens leaves
7. $0 \cdot 31 – 0 \cdot 28$ μm—detrimental effect
8. less than $0 \cdot 28$ μm—in this band the radiation can kill plants.

9.4.2   Leaf Absorption

Radiation cannot penetrate deeply into dense, horizontal-leaf stands of vegetation. For example, with alfalfa of approximately 60 cm (2 ft), about 10% of the radiation is depleted for every 10 cm (4 in) penetration into the crop. It is suggested that the ideal arrangment is where the upper leaves are vertical and allow penetration of the radiation to the lower horizontal leaves. It is this pattern of leaf orientation that is often bred for nowadays, through processes of genetic variability, and of such a pattern is the so-called 'wonder' rice. In addition, the orientation of the rows of the crops is important for, as seen in Chapter 2, the amount of radiation received will be a function of row orientation or angle. When radiation impinges on a leaf, three separate processes occur: some of the radiation is absorbed, some is reflected, and some is transmitted through the leaf to the vegetation or ground below. The percentage of absorption, reflection and transmission is not constant with wavelength, neither is it constant from plant to plant. It has been shown that desert (xerophytic) plants reflect more radiation at all wavelengths than do the mesophytic, or medium moisture-loving, types of plants. This fact enables desert plants to retain a lower temperature under extreme radiation conditions than would otherwise be the case.

9.4.3 Photosynthesis

Photosynthesis is the process of the production of carbohydrates from water and carbon dioxide under the effect of solar radiation (*see* Section 9.11); a process that is the source of practically all organic substances. The radiation of prime importance is in the two bands around 0.45 μm and 0.65 μm. A maximum of about 5% of the solar radiation is used in this process but, in practice, this is reduced to only about 1–2%. The photosynthetic rates of most leaves increase with the intensity of the light falling on them but, at a particular intensity, the leaf becomes light-saturated and the photosynthetic rate is then independent of light intensity. Plants can be classified into two groups—the sun species (for example, rice, sugar cane, wheat, and most field crops) and shade species (oxalis, philodendron). Sun species become light-saturated at about 3000–6000 foot candles, while shade species become saturated at about 1000 foot candles (1 ly min^{-1} is approximately 6000–7000 foot candles). It must be remembered that lower leaves are shaded by the upper ones so that upper leaves may be light-saturated while those below are not. Some plants are thought to need shade for better growth (cocoa and tea) but recent experiments suggest that these actually grow better in the open. Coffee and sunflowers, however, do prefer moderate shading for best production. Since, basically, all dry matter originates from photosynthesis, there is often a simple relation between radiation and crop yield, such as that given by Chang (1968).

9.4.4 Photoperiodism

Photoperiodism is the response of the plant to the 24-hour light pattern cycle. It was first identified in 1920 and it has since been shown that the critical factor is actually the length of the night. Plants are often subdivided into three classes:

1. long day (short night)—these flower when the daylight is greater than 14 hours (winter wheat, barley, spinach, radish, larkspur)
2. neutral—any length of daylight (carrot, summer rice, pansy, tomato, squash)
3. short day (long night)—flower when daylight is less than 10 hours (sweet potato, some soybeans, orchid).

Some plants can change in their response to day length, for example the strawberry has a short day length requirement for floral initiation, but a long day requirement for fruit initiation. Some tropical plants have been found to respond to differences of day length as small as 15 minutes. An interesting example is with a species of cocoa that grows better in a 16-hour day length than in a 12-hour day length, an unusual finding for a plant of the tropics where the day length is always approximately 12 hours.

9.4.5 Phototropism
This is the property wherein the direction of growth of the plant is determined by light and its direction. The active spectral range is from 0·43–0·48 μm. The stems of most plants have positive phototropism, that is, they grow towards the light, but many leaves exhibit transverse phototropism by growing perpendicular to the direction of light. Some cacti arrange themselves so as to receive only the morning and evening sun directly, and these are referred to as compass plants. The sunflower, turning to the rising sun, is another common example of phototropism.

9.5   Plant Response to Temperature

9.5.1 General
Regardless of how favourable radiation and moisture conditions may be, each plant will cease growing if the temperature goes below a certain threshold, T_l, or above another value, T_u. In contrast, the greatest growth rate will occur at an optimum temperature, T_o. The three temperatures T_l, T_o, T_u are known as cardinal temperatures and they may vary both with the plant and its developmental stage (Table 9.1 and Figure 9.1).

Respiration, the process in which carbon dioxide is produced from oxygen, is increased by high night temperature (nycto-temperature)

TABLE 9.1

	T_l	T_o	T_u
Barley, oats, rye, wheat	0–5°C	25–31°C	31–37°C
Melon, sorghum	15–18°C	31–37°C	44–50°C

while high day temperature (photo-temperature), up to a threshold of about 30–37°C (86–98°F), increases the photosynthetic rate, after which the rate decreases. It should be noted that some alpine and arctic plants have their maximum photosynthesis and photosynthetic rate at about only 15°C (59°F).

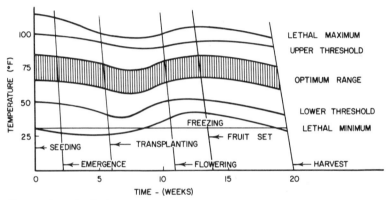

FIGURE 9.1   Cardinal temperatures as a function of development for tomatoes. (Reprinted by permission of William P. Lowry from *Weather and Life*, Academic Press, 1969)

9.5.2   Thermoperiodicity

Thermoperiodicity is the response of the plant to the diurnal variation of temperature, a process rather similar to photoperiodicity. Some plants need a cool night; for instance, for tuber production a 12°C (54°F) nycto-temperature is desirable, while the nycto-temperature is also a dominant factor for chilli peppers, potatoes, and tobacco. In contrast, the photo-temperatures are dominant for peas and strawberries. The pattern of some of these changes, compared with conditions at chosen stations, is shown in Figure 9.2.

9.5.3   Heat Units

This concept, using the idea of degree-days, is about 200 years old and dates from a suggestion of Réaumur. The expression for the number of heat units, H, is given by

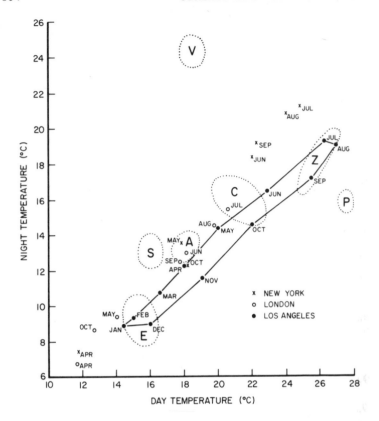

FIGURE 9.2   Thermoperiodicity of some plants and conditions in New York, London and Los Angeles (Kimball and Brooks, 1959)

$$H = \sum_i (\bar{T}_i - T_{th})$$

and details are given in Section 9.11. The value of T_{th} varies from crop to crop and, for example, has been suggested as 4°C (40°F) for peas, 10°C (50°F) for citrus. The disadvantages in the use of such a simple concept are as follows: the expression assumes a linear relationship with temperature, a supposition that is not generally confirmed in practice; secondly, T_{th} can, and often does, change with the stage of development of the plant; thirdly, the daily range of temperature is omitted from such calculations; fourthly, when the temperature is very high it can affect the crop or vegetation in a detrimental manner, but H would receive a large number of units, thereby wrongly indicating an improvement.

Another problem is that two distinctly different temperature patterns can give rise to exactly the same number of heat units, a condition that is shown in Figure 9.3.

Some investigators have used the concept of a photothermal unit, in which the temperature element is multiplied by the day length, and recently the concept of radiation units has been suggested.

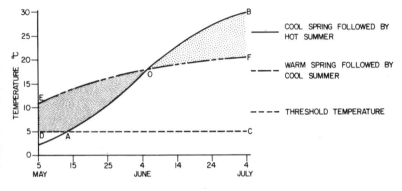

FIGURE 9.3   Schematic diagram of heat summation. (Reprinted by permission of Inland Printing Company from *Agricultural Meteorology* by Jen-Yu Wang, 1963)

9.6   Plant Response to Moisture

9.6.1   Soil Moisture

Full or potential transpiration from vegetation needs adequate available soil moisture. This means that in most cases full transpiration is unlikely to be taking place. Less than 1% of the water passing from the plant is used in photosynthesis, and generally the rate of photosynthesis decreases rapidly after 30% of the water content of the leaves is lost, and ceases altogether at about a 60% loss. Moisture is generally available to the plant between the thresholds of wilting point and field capacity, but some crops have the ability to extract moisture from the soil when other crops are under pronounced moisture stress. Some studies have shown a very good correlation between crop yields and evapotranspiration. Such a relationship would indicate the plant dependence upon available soil moisture.

9.6.2   Air Moisture

A combination of high radiation and low relative humidity can cause stress in plants growing in wet soil when the water uptake rate is less than the transpiration loss. A low relative humidity, or a high saturation deficit, will give a large driving force for transpiration. Higher relative humidity will often reduce evapotranspiration and thereby benefit the crop.

Dew, the condensation of moisture on vegetation generally occurring at night, can be of two kinds—either dew fall, where the condensation is taking place from the air above the vegetation, or dew distillation, when moisture is moving from the soil and condensing on the cool leaves of the plant. Dew is normally at a maximum in the sub-tropical humid areas and can amount to about 75 mm (3 in) per year with a daily maximum around 0·5–1 mm (0·02–0·04 in). Dew can be beneficial by bringing moisture to the plant at a correct time, but it can also be detrimental by giving an environment perfect for the transmission of rust or blight.

9.6.3   Drought

Drought is an extremely difficult term to define, for it has special, specific connotations for the meteorologist, the agriculturist, the hydrologist and the economist. Generally, drought is expressed in terms of the weather and climate of a particular region and is often referred to as a period of abnormally dry weather sufficiently prolonged for the lack of water to cause a serious hydrologic imbalance in the affected area. The vegetation and the animals of an area are usually adapted to the average conditions of that region, and they will suffer severely when the deviation in precipitation is both intense and prolonged. The size of the area must also be taken into consideration, for a simple example will show it is a fundamental feature of the definition. If, for the purposes of the illustration, it is considered that a year is a drought year when the rainfall is less than 75% of the average (this is a value that is often used in domestic animal studies and pasture considerations) then during the past eighty years the United States of America (as a whole) has suffered only one drought, Texas (as a whole) has had six, the Edwards Plateau (in Texas) fifteen, while San Angelo (on the Edwards Plateau) has experienced twenty-one droughts.

It must also be realised that in some regions a drought is normal! By this is meant the following: if a month with less than 75% of its annual rainfall is accepted as a drought month, then it can be shown for San Angelo that 53% of the months would fall into the drought category, whilst only 27% have in excess of 125% of the normal amount; therefore, it is 'normal' for a drought month to be recorded.

Detailed investigations have not shown a pronounced or usable cycle with drought occurrences. This point will be seen in Table 9.2 where drought years for various parts of Texas are given. In only one year did drought occur in all ten climatic regions of the state. Another interesting aspect is that there appears to be no persistence between the

TABLE 9.2

Drought years in the ten climatic divisions of Texas (values expressed as a percentage of normal rainfall in the Division) (Texas Almanac, 1974–75)

Year	High Plains	Low Rolling Plains	North Central Texas	East Texas	Trans-Pecos	Edwards Plateau	South Central	Upper Coast	Southern	Lower Valley
1892					68			73		
1893			67	70		49	56	64	53	59
1894					68					
1897							73		72	
1898									69	51
1901		71	70			60	62	70	44	
1902									65	73
1907										65
1909			72	68	67	74	70			
1910	59	59	64	69	43	65	69	74	59	
1911										70
1916		73		74	70		73	69		
1917	58	50	63	59	44	46	42	50	32	48
1920										71
1921					72					73
1922					68					
1924			73	73		71		72		
1925			72				72			
1927								74		74
1933	72				62	68				
1934	66				46	69				
1937									72	
1939							69			72
1943			72							
1948			73	74	62		73	67		
1950							68		74	64
1951					61	53				
1952	68	66			73				56	70
1953	69				49	73				
1954	70	71	68	73		50	57		71	
1956	51	57	61	68	44	43	55	62	53	53
1962						68			67	69
1963			63	68		65	61	73		
1964	74				69					63
1970	65	63				72				

1931–60 normal rainfall for the above divisions respectively (inches)—
18·51, 22·99, 32·93, 45·96, 12·03, 25·91, 33·24, 46·19, 22·33, 24·27.

rainfall amount in one month and the rainfall amount in the following month, in other words persistence is zero. There is a tendency to suggest that, in a particular local region, droughts are a cyclic phenomenon. The facts, however, indicate that droughts re-occur; they do not recur.

9.7   Plant Response to Combination of Elements

1. With vegetation there is a point at which the respiration rate reaches the photosynthetic rate. This is referred to as the compensation point; above this point there is no gain in dry matter for the plant and this compensation point is found to be a function of both light and temperature. For example, rice has a compensation point at 4°C (40°F) and 150 foot candles and at 27°C (80°F) the value has increased to 1400 foot candles. At the lower temperatures the average compensation point is around 100–150 foot candles for the sun-loving species and as low as 50 foot candles for shade species.

2. Photosynthesis is found to increase not only with light intensity but also with temperature. In addition, an increase in carbon dioxide concentration will lead to an increase in the potential photosynthesis.

3. The transpiration rate from a plant is a function of a combination of radiation, temperature, relative humidity and wind. This aspect is so important that it is detailed in the next section.

9.8   Evaporation and Evapotranspiration

Evaporation is defined as the loss of water from either soil or water as it changes from a liquid to a vapour state in the air. When live vegetation is also involved, then an extra process, transpiration, contributes a further water loss. The total loss for a vegetative cover is called evapotranspiration. Many attempts have been made to assess the magnitude of this water loss but the work is complicated by the fact that there are really seven distinct problems inherent in the process.

1. Loss from an open water surface—evaporation, E_o
2. Loss from a soil surface with adequate moisture—potential evaporation, PE_s
3. Loss from a soil surface under natural conditions—actual evaporation, AE_s
4. Loss from vegetation with adequate moisture—potential transpiration, PT

5. Loss from vegetation under natural conditions—actual transpiration, AT
6. Loss from a soil/vegetation complex with adequate moisture—potential evapotranspiration, PET
7. Loss from a soil/vegetation complex under natural conditions—actual evapotranspiration, AET

From a practical viewpoint (4) and (5) are rarely considered as separate components because the vegetation is almost invariably in a soil medium and the problem becomes concentrated upon (6) and (7).

In meteorology interest is focused upon (1) for, with the need for a standard surface, water presents a more uniform medium than soil or vegetation. Measurement of E_o are made with either open water pans (evaporimeters) or porous surface models (atmometers). There are many varieties of evaporimeters in use (Platt and Griffiths, 1972, W.M.O., 1968) and because of their different sizes and methods of exposure there is not a perfect agreement among them. In fact, the correlation decreases as the difference between the sizes of the pans increases. Many attempts have been made to calculate E_o using either physical relationships or empirical equations, but unless the latter have a high correlation coefficient, apply over a large area, and are simple to use, their efficiency is debatable. Atmometers are simple, portable and of low cost but their results, especially when the instrument is exposed in the instrument shelter, show poor correlation with E_o and PET.

The methods of calculating evapotranspiration or evaporation fall into three classes—the aerodynamic method, the energy balance approach and empirical equations. The aerodynamic method, sometimes known as the Dalton or mass transfer method, employs an equation of the form

$$E = N f(u) (e_s - e_d)$$

where N is a value (mass transfer coefficient) that depends not only on the surface but also on where the corresponding meteorological measurements are made, u is the free air wind speed, e_d is the saturation vapour pressure at the dew point temperature and e_s is the vapour pressure at the surface. The method can be used to find E_o, PE_s or PET when the radiative component is ignored.

The energy balance approach assumes that the advected energy, the heat flux to the soil, heat storage in the crop and the energy of photosynthesis are zero—not all of these assumptions can be justified under all conditions. Then the energy budget equation takes the simplest form

$$R_n = A + E$$

where R_n is the net radiation, A is the heat flux to the air and E is the energy used in the evaporation (evapotranspiration) process. The ratio between A and E is called the Bowen ratio, an expression that is extremely difficult to calculate with accuracy. To avoid some of these problems and to eliminate the difficulty of obtaining the temperature of the evaporating surface Penman combined both techniques and presented the formula

$$E = cR_n + (1 = c)E_a$$

The derivation is given in Section 9.11, where $c = \Delta/(\Delta + \gamma)$, $\Delta = $ the slope of the saturation vapour pressure v. temperature curve at the air temperature, T_a, (de_{sat}/dT_a), and γ is the psychrometric constant. Because of the shortage of measurements of R_n, Penman substituted empirical relationships using sunshine duration, humidity and temperature. As T_a increases so does c, at 0°C (32°F) c is 0·4, at 6°C (43°F) it is 0·5, at 13°C (55°F) 0·6, at 21°C (70°F) 0·7 and at 31°C (88°F) 0·8. This means that at higher temperatures the energy term dominates the aerodynamic term.

When Penman used this equation to find PET, as distinct from E_0, he defined the vegetation cover as being a short green crop of uniform height. A tall crop will change radically the radiation pattern and the air flow, resulting in greater potential evapotranspiration. In one test, a crop of lucerne 40 cm (16 in) taller than the control plot experienced over twice the potential evaporation of the shorter crop. Penman found that PET/E_0 varied from about 0·6–0·8, being greater during the longer days of summer. However, his method for calculating PET takes no note of the fact that in the process of transpiration there is a resistance to the diffusivity of water through the stomata and the cuticle, such resistances do not occur in E_0.

Empirical relationships for calculating potential evapotranspiration are numerous, ranging from those using only meteorological variables, generally temperature, to those that include various crop factors, and the interested reader is referred to Chang (1968) or W.M.O. (1968).

Actual evaporation or evapotranspiration presents a much more difficult problem. Measurements of this can be made only by the use of weighing lysimeters, which are rather expensive equipment and need to be capable of weighing with extreme accuracy—for example, one model, in which the enclosed soil and vegetation totalled about 3000 kg had an accuracy of ten grams, or one part in 300 000. A lysimeter is simply a buried tank, containing soil and the vegetation to be studied. The instrument should be large enough to reduce the boundary effects, it should be exposed amid similar conditions of soil and vegetation and care should be taken to ensure comparable water and thermal regimes between the lysimeter and the natural conditions.

When the water in the soil is below the field capacity many authors suggest that evaporation cannot proceed at the potential rate. The relationship between AE_s and PE_s has been the subject of much discussion and Figure 9.4 shows some of the suggested distributions as

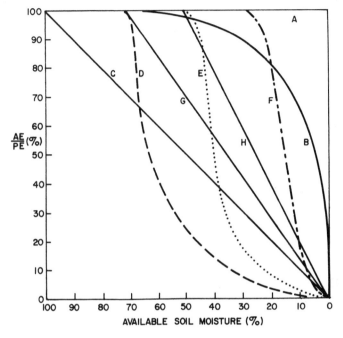

FIGURE 9.4   Relationships of actual to potential evaporation with varying available soil moisture. (Reprinted with permission of Canada Department of Agriculture from *Tech. Bull.* 78, *Agrometeorol. Sect.*, by W. Baier *et al.*, 1972)

a function of the soil moisture tension. One point on which there is reasonable agreement is that there is a threshold level, perhaps dependent upon the soil, above which the evaporation is controlled, mainly by atmospheric conditions, and below which it is the soil characteristics that determine evaporation (Derendinger, 1971). For calculations of *PET* or *AET*, the depth of the rooting system must also be considered.

9.9 Energy Budgets

As mentioned earlier, it is energy fluxes and balances that determine the conditions under which the plant develops. If a plant is transpiring, this will act to cool the surfaces, in much the same way as when animals perspire. In Table 9.3 leaf temperatures are given for certain types of conditions and the table indicates what an important role is played by convection and transpiration, especially at high temperatures. It is

TABLE 9.3
Leaf temperatures (°C) for conditions listed and an
air temperature of 30°C (Gates, 1972)

R (cal cm^{-2} min^{-1})	Radiation only	Radiation and convection V (cm s^{-1})			Radiation, convection, and transpiration V (cm s^{-1})		
		10	100	500	10	100	500
0·6	20	28	29	30	25	27	29
1·0	60	41	34	32	33	31	30
1·4	89	53	39	34	40	35	32

interesting to note that with a medium solar load (1 ly min^{-1}) and a light wind of 1 m s^{-1} (2·2 mph) the leaf temperature will be only 1 deg C (18 deg F) above the ambient air temperature. Of course, the high radiation values are reached generally only on leaves near the top of the plant. In the extreme case of high radiation and low wind speed, it is seen that convection effects a cooling of 36 deg C (65 deg F) and transpiration 13 deg C (23 deg F).

9.10 Domestic Animals

While it is generally realised that most domestic animals respond to climate and weather, the greater percentage of specific studies have been concerned with dairy cattle, for they are particularly sensitive to atmospheric changes (Figure 9.5). Temperatures above about 35°C are generally detrimental, and, if associated with high radiation and/or high humidity, can cause serious stress. Many experiments have been conducted in hot areas to provide inexpensive shelters, but natural shading from trees proves to be best. Beef cattle have a much greater tolerance to temperature, but the effect of high temperature, acting through pasture and water, can be to reduce the growth rate. Zebu, the

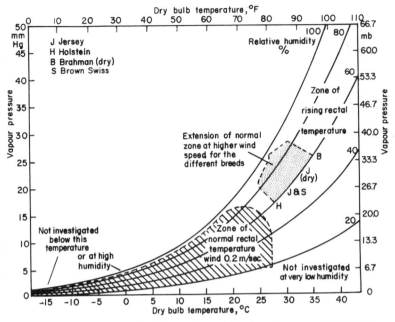

FIGURE 9.5 Comfort zone for dairy cattle as relation to environmental temperature, humidity and air movement. (Reprinted with permission of Missouri Agricultural Experiment Station from Bulletin 552, 1954 by H. H. Kibler and S. Brody)

hump-backed cattle of the tropics, have a greater heat tolerance than the European cattle and cross-breeding has resulted in some breeds that can accommodate well to high heat and humidity.

Sheep are little troubled by cold conditions for their fleece can protect to even −40°C (−40°F), but if snow packing occurs on the fleece the insulation is drastically reduced and death can result rapidly. Newly born lambs are particularly susceptible to the cold, while recently shorn sheep should be protected against snow or rain. All domestic animals should have protection from severe conditions, such as blizzards and hail.

9.11 Appendix

1. Photosynthesis

The conversion of radiant energy to chemical energy is accomplished by photosynthesis. In the process carbon dioxide diffuses into the leaves, through the stomata, and dissolves in the water of the cell walls. The reaction may be expressed as

$$6CO_2 + 6H_2O + \text{solar energy} = C_6H_{12}O_6 + 6O_2$$

carbon dioxide   water   sugar
                                                (hexose)

2. Degree–days

In this accumulation, $\bar{T}_i$ is the mean daily temperature on day i, A is an arbitrary temperature and t is the period in days covering all the days 'i' to be studied. Then, DD, the total number of degree-days is

$$\sum_{i=1}^{t} (\bar{T}_i - A).$$

Whenever $\bar{T}_i$ is less than A, the difference counts as zero. For example, if $A = 60$ and during a week the mean daily temperatures were (in °F) 75, 68, 61, 60, 57, 50, 64, then

$$DD = (15) + (8) + (1) + (0) + 0 + 0 + (4) = 28.$$

3. The energy budget method (Penman)

From page 109 we have seen that

$$E = (e_s - e_d)f(u) \tag{1}$$

and

$$R_n = E + A = E(1 + \beta) \tag{2}$$

where β, the Bowen ratio, is given by

$$\beta = \gamma \frac{(T_s - T_a)}{(e_s - e_d)} \tag{3}$$

γ being the standard psychrometric constant.

From (1) we may substitute e_a (the saturation vapour pressure at air temperature, T_a) for e_s, to obtain E_a, defined by

$$E_a = (e_a - e_d)f(u) \tag{4}$$

From (1) and (4)

$$(e_s - e_d) = (e_s - e_a)\left(1 - \frac{E_a}{E}\right)^{-1}$$

so that

$$\frac{T_s - T_a}{e_s - e_d} = \frac{T_s - T_a}{e_s - e_a}\left(1 - \frac{E_a}{E}\right) = \Delta\left(1 - \frac{E_a}{E}\right) \tag{5}$$

where Δ is as given on page 110.

Therefore, from (2), (3) and (5)

$$E = R_{\text{n}} / \left[1 + \frac{\gamma}{\Delta} \left(1 - \frac{E_{\text{a}}}{E} \right) \right] = R_{\text{n}} E / \left[E \left(1 + \frac{\gamma}{\Delta} \right) - \frac{\gamma}{\Delta} E_{\text{a}} \right]$$

or

$$E = \left(R_{\text{n}} + \frac{\gamma}{\Delta} E_{\text{a}} \right) / \left(1 + \frac{\gamma}{\Delta} \right) = \frac{\Delta R_{\text{n}} + \gamma E_{\text{a}}}{\Delta + \gamma} = c R_{\text{n}} + (1 - c) E_{\text{a}}$$

where $c = \dfrac{\Delta}{\Delta + \gamma}$.

References

Budyko, M. I. (1968) *Solar Radiation and the Use of it by Plants, in Agro-climatological Methods*, U.N.E.S.C.O., 38–53

Chang, Jen-Hu, (1968) *Climate and Agriculture*, Aldine-Atherton, Chicago, 304 pp.

Derendinger, G. L. (1971) 'The relationship of evaporation in the falling rate stage to water diffusivity in soils', Texas A & M University, Dissertation, 103 pp.

Gates, D. M. (1972) *Man and His Environment: Climate*, Harper and Row, New York, 175 pp.

Kimball, M. M. and Brooks, F. A. (1959) 'Plant climate of California', *Calif. Agric.*, **13**(5), 7–12

Lowry, W. P. (1969) *Weather and Life*, Academic Press, New York, 305 pp.

Mather, J. R. (1974) *Climatology: Fundamentals and Applications*, McGraw-Hill, 412 pp.

Platt, R. B. and Griffiths, J. F. (1972) *Environmental Measurement and Interpretation*, Krieger, Huntington, New York, 235 pp.

Texas Almanac (1974–75) *The Dallas Morning News*, 705 pp.

Wang, Jen-Yu. (1963) *Agriculture Meteorology*, Pacemaker Press, Milwaukee, 693 pp.

Williams, C. N. and Joseph, K. T. (1973) *Climate, Soil and Crop Production in the Humid Tropics*, Oxford University Press, Kuala Lumpur, 177 pp.

World Meteorological Organization (1968) *Practical Soil Moisture Problems in Agriculture*, Tech. Note 97, 69 pp.

Suggested Reading

Molga, M. (1962) *Agricultural Meteorology*, Office of Technical Services, U.S. Dept. of Commerce (trans. from Polish), 253 pp.

U.N.E.S.C.O. (1968) *Agroclimatological Methods, Natural Resources Research*, 392 pp.

Walter, H. (1973) *Vegetation of the Earth*, Heidelberg Science Library, Vol. 15, 237 pp.

World Meteorological Organization (1970) *Meteorological Observations in Animal Experiments*, Tech. Note 107, 37 pp.

World Meteorological Organization (1970) *Weather and Animal Diseases*, 49 pp.

10

Atmospheric Modification

10.1 Introduction

Man usually is not satisfied with the climate that he experiences; a period of a few hot days elicits the comment 'Oh, for a cool spell', a time of showers brings out the longing for a sunny day. The conscious modification of Man's micro-habitat began with the wearing of clothing and the control of fire to warm the cave. This system has been continued until the present day with more sophisticated heating and, relatively recently, cooling techniques. However, such methods deal with enclosed spaces and therefore need relatively little energy to retain a desired set of conditions. Outside in the free atmosphere, the problem is very different.

10.2 Irrigation

This is perhaps the earliest way in which man altered the climatic conditions of an area. Nowadays an estimated $1700 \text{ km}^3 \text{ y}^{-1}$ is involved in non-returned irrigation water (Lvovich, 1969). An irrigated area has a local temperature lower than the surrounding region due to increased evaporation, and also a higher relative humidity. Perhaps the most extreme examples of the effect of a water body (natural) and irrigation are to be found in the oases of the sub-tropical deserts. In a location in southern Arabia the author noted a change from 48°C (118°F), 12% r.h. outside to 30°C (103°F), 31% at the centre of only a small oasis, about 50–70 m in radius. Thus, locally, irrigation brings about a temperature decrease.

However, because irrigation causes vegetation to remain green, the global temperature is raised, due of the decrease in the reflection of incoming solar radiation, and Budyko (1971) estimates that the earth's mean surface temperature has increased 0·07 deg C (0·13 deg F) because of this effect.

10.3 Shelter Belts

Windbreaks are perhaps the earliest example of deliberate weather modification. They have proved their worth in the control of snow drifting, soil erosion, and in reducing wind pressure on objects in their lee. Their effectiveness is dependent, naturally, on the degree of constancy of the 'undesirable' wind: if it is very variable the shelter belt may well prove more of a liability than a benefit.

A shelter belt will have the effect of reducing the speed downwind, roughly in the manner shown in Figure 10.1. Close to the break, both to leeward and windward, small independent eddy cells can be generated

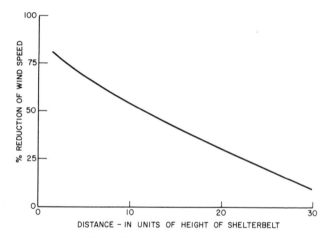

FIGURE 10.1   Effect of shelter belt in reducing wind speed

so that wind direction experiences a complete reversal. Jensen (1954) has shown that the most efficient windbreaks are those with about a 40–50% coverage in the vertical—not those with a complete barrier, for the effect of these is not noted far downward.

A windbreak naturally modifies the conditions with respect to radiation, and thereby air and soil temperatures, and precipitation. Perhaps the most practically significant induced change is in evaporation and transpiration. The rate of both these phenomena is reduced at lower wind speeds. The changes induced in the local micro-climate are shown in Figure 10.2, taken from a useful publication by Read (1965). This paper also indicated that in a zone $1\frac{1}{2}$ to 12 times the height of the barrier, downwind yield can be increased up to 50% (maximising at 4–8H downwind). Immediately adjacent to the barrier, the yield decreased due, among other factors, to the competition

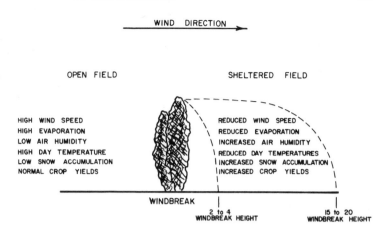

FIGURE 10.2   Local climate of open fields compared with that of fields affected by windbreaks (after Read, 1965)

between trees and crops for available soil moisture. Snow fences, made of open, slatted boards, are often used to ensure a more uniform snow cover for crop protection and a resultant soil moisture for early planting after the thaw. Another important aspect of a shelter belt is in the influence on dew deposition—the amount within 2 to 3 times the height of the break being about twice that in the open (Steubing, 1952).

10.4   Mulches

In practice, mulches are materials used for the reduction of water loss. They alter the chemical and physical characteristics of the barrier layer between the air and the soil. The changes are not related simply to moisture but involve radiation, temperature and wind also. Sometimes the mulches are of natural materials, such as snow, sod or other plants; often they are artificial—paper, sawdust, plastics, aluminium sheet and others. Various coloured substances have been used to increase the soil temperatures, and an interesting study in the USSR noted that cotton ripening was achieved a month earlier with the application of 100 lb of coal dust per acre (100 k ha^{-1}).

An aspect of atmospheric modification that can be considered in this section is the use of retardants (chemical compounds) to reduce evaporation. These have been applied mainly to open water surfaces, but occasionally have been used on plants and soil. The fatty cetyl alcohols, which are non-toxic, have been used as monomolecular layer films on reservoirs and lakes. However, the small surface tension that holds the film in place is easily overcome by wind pressure and friction, once the speed exceeds about 2 m s⁻¹ (4·4 mph), and the water is uncovered allowing evaporation to proceed unhindered. In the hot, dry areas evaporation can exceed 1 cm day⁻¹ (0·4 in day ⁻¹), reaching as much as 2–4 cm day ⁻¹ (0·8–1·6 in day⁻¹) in the region of Lake Nasser, the Aswan Dam, in Egypt.

10.5 Artificial Stimulation of Rain, and Hail Suppression

It was about a quarter of a century ago that Vincent Schaefer demonstrated that the precipitation process could be initiated in a cloud by seeding it with solid CO_2 (dry ice). Such a potential 'control of the weather' caught the imagination of both the scientific community and the general public, but it was not until 1957 that a Presidential Committee concluded that on the basis of statistical evaluation, cloud seeding in the mountainous areas of the western United States 'produced an average increase of 10–15%—with a satisfactory degree of probability that the increase was not the result of the natural variations in the amount of precipitation' (Nat. Acad. Sci., 1973). In a progress report in the same publication, this statement is confirmed and some other aspects considered:

1. that cloud seeding can cause rain decreases as well as increases for certain types of clouds;
2. experiments to diminish the intensity of hurricanes have been started, with preliminary results that provide some basis for optimism;
3. a major programme to explore the possibility of decreasing the damage from hail has been started in the State of Colorado.

Cloud seeding can be carried out from ground generators or rockets, or by use of aircraft. The type of nucleation used is generally silver iodide, but in warm clouds (most above 0°C) finely powdered, dry common salt (NaCl) is often used. Some of the most prolonged and statistically rigorous experiments have been carried out in Israel (Neumann et al., 1967) with results that indicate an increase of 18% during seeding over the entire area, reaching a maximum of 27% within the centres of the target areas.

The author was involved in some tests in East Africa during the 1960s and, although the trials were not statistically designed, reached the opinion that clouds about to precipitate could be induced to rain from 10–30 minutes earlier when seeded with dry salt. This could bring rainfall to the plains where agriculture was practised, as distinct from having rain on nearby mountain slopes.

It was in 1969 that Project Stormfury conducted seeding trials on Hurricane Debbie and, although scientific opinions differ on the impact of the seeding by silver iodide, the evidence does offer hope that hurricane intensity may be reduced by this method (Gentry, 1970). Much further work remains to be done. Some of the major studies in hail modification have been carried out in the Soviet Union where they cover some millions of acres of farmland. Fedorov (1967) has noted that hail damage in the protected area was 3–5 times smaller than in the unprotected area, while the cost of the protection was only 2–3% of the value of the crops involved.

10.6   Frost Protection

Freeze conditions occur in one of three ways: cold air may be brought into the region (advection); it may be produced locally by rapid natural cooling (radiation) or it may be the outcome of a combination of the two. Generally, when cold air is advected into an area, the methods of protection are rarely efficient, since too much energy must be supplied to keep the air mass above the critical threshold temperature.

There are five basic means of protection. Covering the plant is a simple method and generally can work well. Examples using this technique are the flooding of cranberry bogs, soil covering of young plants, and straw covering of vines. The material used must be opaque to the long wave radiation. A new method uses a non-toxic,`protein-based fire fighting foam to cover row crops, such as tomatoes and strawberries. Blankets of foam (2–7 cm thick) applied in the late afternoon were effective against radiative and advective cooling even when temperatures fell below $-4°C$ (25°F) for six hours. The foam is easily applied, and disperses spontaneously or is dissipated by water or rain (Siminovitch et al., 1972).

A fog will act in much the same way as the covering method and this is the reason for the popularity of smoke generators and smudge pots. These 'oil' fogs, however, do not produce particles sufficiently large to give real protection.

Wind machines have been shown to give greater success in some areas. During radiation frost conditions, a shallow inversion layer is often found such that 15 m (50 ft) above the surface the air may be 6–8 deg C (10–15 deg F) warmer than at the ground. Mixing the air with fans (static or helicopters) can therefore bring about a slight heating.

The sprinkling technique uses the physical property of water that when it changes from liquid to ice at 0°C, there is a release of 80 cal g^{-1} (the latent heat of fusion). This can then prevent the leaf temperature from falling below 0°C, provided a film of water is maintained. The technique will only keep the watered parts at 0°C, at best, and may result in such a build-up of ice that the plant cannot support itself. It is clear that the rate of sprinkling, and droplet size, are of major importance. In addition, the added water can reduce the capacity of the soil to support mechanised equipment, increase leaching and decrease bacterial action in the soil. Sprinkling can only be used efficiently when the wind speed is below about 5 m s^{-1} (11 mph) and the relative humidity is above 60%.

The heating method of frost protection is the most favoured, with the heaters burning coal, oil, wood, solid fuels or other fuels. A large number of small heaters is preferable to a few large heaters, for in the latter case the powerful thermals may break through the inversion and draw in cold air, thus worsening the situation. Experiments with solid fuel burners suggest that three or four heaters per tree can maintain a 3–7 deg C increase, and a 2–4 deg C increase even in winds of 2–5 m s^{-1} (5–11 mph).

10.7   Aircraft Modifications

Since man has been flying in the upper troposphere and stratosphere, there have been many warnings concerning the possible changes that might occur in the amount of high cloud. Such changes, generally assumed to be increases, would naturally bring about reductions in solar radiation receipts, with corresponding, and unknown, changes in the climate of certain areas. The discussion centred around the problems introduced by the supersonic transports (SSTs) has to remain at an academic or mathematical level, although a publication of the National Academy of Sciences has concluded that no climatic effects could be expected.

A study of high cloud patterns as many stations both before and after the advent of jet aircraft has shown interesting, but sometimes confusing, results (Griffiths and Harriman, 1974). In general, those regions over which jet traffic is heavy, have shown increases in high cloud while, with some unexplained exceptions, other regions have shown no significant change. However, a study of solar radiation amounts at the same stations gave no significant alterations.

10.8 Vegetation Effects on Precipitation

There have been many reports concerning the effect, if any, of vegetation on precipitation. The effects of vegetation upon precipitation may result in a change in actual precipitation or only in the measured precipitation.

Of the four types of rainfall (convective, orographic, convergent, and cyclonic), it is generally agreed that only the first two are ever affected, while convergent and cyclonic precipitation are never altered by vegetation. Some published data suggest that vegetation, mainly forests, causes from 3 to 20% increase in convective precipitation over that in cleared areas. Orographic precipitation may be increased from 2 to 10% due to the added height. Vegetation may cause lengthening of the rainy season, and the precipitation increase may be greater during the dry years than the wet ones. Part of the increase in measured precipitation may be due to the collection of fog in the gauges. Studies suggest that forests increase fog drip from 3 to 50% over areas that have no trees.

Other data suggest that vegetation has no effect whatsoever on the quantity, duration or intensity of precipitation. Some vegetation types may even cause a slight decrease in precipitation because of their cooling effect on the air.

It is readily obvious that at present no positive proof is available to permit a decision concerning actual effects of vegetation upon precipitation. It is a question which requires more controlled work and research.

References

Budyko, M. I. (1971) *Climate and Life*, Hydroglobal Publishing House, Leningrad, 508 pp.

Fedorov, Ye. K. (1967) 'The artificial modification of meteorological processes', W. M. O., Document CC–VI, 63, Appendix C.

Gentry, R. C. (1970) 'Hurricane Debbie modification experiments, August 1969', *Science*, **168**, 473–5

Griffiths, J. F. and Harriman, R. E. (1974) 'High cloud increases in recent years', *Int. J. Environ. Stud.* **5**, 271–3

Jensen, M. (1954) *Shelter Effect*, Danish Technical Press, Copenhagen, 264 pp.

Lvovich, M. I. (1969) *Vednie Resoursi Budushevo (Water Resources of the Future)*, Prosveschenie, Moscow, 174 pp.

National Academy of Sciences (1973) *Weather and Climate Modification, Problems and Progress*, 258 pp.

Neumann, J., Gabriel, K. R. and Gagin, A. (1967) 'Cloud seeding and cloud physics in Israel: results and problems', Proc. of Int. Conf. on Water for Peace, Washington, D.C.

Read, R. A. (1965) *Windbreaks for the Central Great Plains*, Lincoln, Nebraska, 7 pp.

Siminovitch, D., Rheaume, B., Lyall, L. H. and Butler, J. (1972), *Foam for Frost Protection of Crops*, Canada Dept. of Agric., Ottawa, Publication 1490

Steubing, L. (1952) 'Der Tau und Seine Beeinflussung durch Windschutzanlagen (Dew and its influence by shelterbelts)', *Biol. Zentr.*, **71**(5–6), 282–313

11

City Climates

11.1 Introduction

The city has two distinct facets that differ from the rural environment: (a) a compact grouping of buildings and streets, and (b) a concentration of inhabitants whose activities generate a large source of heat. Lowry (1967) has suggested that these two facets account for five basic influences that differentiate between a city and its surrounds.

1. Surface materials—the stone and rock-like materials have a heat conductivity about three times that of wet, sandy soil. Therefore, the city materials will be able to accept and store more energy in a shorter time.
2. Shapes and orientations of surfaces—the variety of these is much greater in a city than in the natural countryside. Figure 11.1 illustrates the way in which incoming energy is reflected in a city in such a manner as to use almost the entire surface to receive and store heat, while in an open countryside the heat-receiving area tends to be in the upper vegetation layer. Also, the many buildings in a city tend to change the air flow pattern, acting as a brake and increasing turbulence while, at the same time, causing channelling of air and creating both 'windy canyons' and 'calm pockets'.
3. Heat sources—the city, especially in winter in temperature zones, can generate a large amount of heat, amounting to an appreciable percentage of the total energy, natural (sun and sky) and artificial.
4. Moisture sources—in a city, drainpipes, gutters and sewers rapidly redistribute the rain; some areas around a house can get a concentration of many times the normal. Similarly, snow is redistributed. In the country more precipitation is retained evenly to be available for evaporation, a process that tends to cool the rural air.
5. Air quality—the city air has a heavy load of contaminants. The implications of this are discussed in Section 11.9.

Of course, each city gives rise to its own (urban–rural) differences but Table 11.1 gives some representative values derived by Landsberg (1970) as a composite of numerous cases. The city complex acts as a unit to change the environment in many ways, both from the

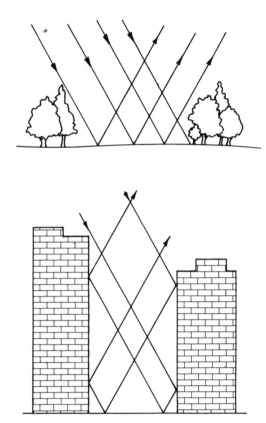

FIGURE 11.1   Incoming energy patterns in rural and urban environments

atmospheric and hydrologic view point. A recent bibliography (Griffiths and Griffiths, 1974) lists many hundreds of publications dealing with this facet of growing interest and implication.

11.2   Radiation

The air of the city normally contains far more dust and other contaminants than rural air. In fact, this pollution can often concentrate as a dust dome over large cities (Figure 11.2). This dome then acts to reduce radiation, especially in the shorter wave (ultraviolet) band, for the particles reflect, scatter and absorb the incident beams. Studies in Vienna, Leipzig and Frankfurt show that this reduction (as a

TABLE 11.1

Average changes in climatic elements caused by urbanisation (Landsberg, 1970)

Element	Parameter	Urban compared with rural
Radiation	On horizontal surface	− 15%
	Ultraviolet	− 30% (winter); − 5% (summer)
Temperature	Annual mean	+ 0·7°C
	Winter maximum	+ 1·5°C
	Length of freeze-free season	+ 2 to 3 weeks (possible)
Wind speed	Annual mean	− 20 to − 30%
	Extreme gusts	− 10 to − 20%
	Frequency of calms	+ 5 to − 20%
Humidity	Relative—Annual mean	− 6%
	Seasonal mean	− 2% (winter); − 8% (summer)
Cloudiness	Cloud frequency and amount	+ 5 to 10%
	Fogs	+ 100% (winter); + 30% (summer)
Precipitation	Amounts	+ 5 to 10%
	Days (with less than 0·2 in)	+ 10%
	Snow days	− 14%

percentage) changes not only with season but even more with solar altitude, ranging from 29–36% (summer-winter) at 10° solar altitude to 15–21% (summer-winter) at 30° and 11% (spring) at 45°. The smoke pollution also causes a much paler blue shade of the sky to be observed.

Sunshine is also generally reduced, by as much as 40–50%, in cities, again with a greater reduction in winter (low sun) than in summer (high sun). With the recent clean air regulations, London has shown a return to better conditions.

A paper by Terjung and Louie (1973) gives an attempt at modelling by computer the urban absorption of solar radiation. Their study shows that downtown areas often absorb less energy than the periphery of the inner zone and that the mere existence of a city is sufficient to generate daytime heat islands.

11.3 Temperature

For many centuries, people have been aware of the thermal differences between a city and its surrounds, Howard's survey *The Climate of London* (1883) being a classic text. More recently Sundborg (1951) has attempted to express the difference between urban temperature and rural temperature, called dT, as a linear function of cloudiness, wind speed, temperature and absolute humidity. For the city of Uppsala, Sweden, Sundborg (1951) derived two equations, one for

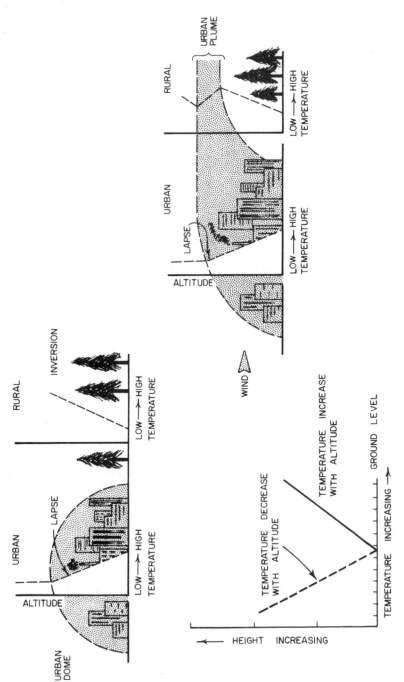

FIGURE 11.2 Comparison of lapse rates in rural and urban environments under different atmospheric conditions (a) on a still, clear night, lapse rate conditions occur in the city and inversions in rural areas, (b) when a regional wind is blowing, the rural lapse rate is modified by the urban plume. Inset shows characteristics of temperature graphs. (Reprinted by permission of John Wiley and Sons from *Climate and Man's Environment*, J. L. F. Oke, 1978.)

day (a maximum dT of 1·4 deg C) and one for night (a maximum dT of 2·8 deg C). Chandler (1965) also showed that dT for London was much greater for mean minimum temperatures (1·9 deg C) than for mean maximums (0·6 deg C). The effects of wind speed (v) and cloud (n) dominate the night time conditions and dT can then be expressed in the form $(a-bn)/v$.

A few hours after sunset on calm, clear days dT will be a maximum value, since heat is still retained in the buildings while rapid cooling is taking place in the open; this difference may reach 10 deg C or more. Mean annual values of dT tend to increase with the size of the city, being proportional to the fourth or fifth root of the population.

Generally there are fewer days with minima below freezing in a city, and a related decrease in the length of the freeze period. In addition, heating degree days (Section 9.11) will be less. All are manifestations of the urban heat island effect, an effect that can be destroyed when wind speeds are high—the critical speed, V, is dependent on city population, P, according to the equation

$$V = 0·5 \, (P)^{0·2}$$

with V in m s^{-1}. The relationship is only approximate, but is generally usable and resembles that for expressing mean dT values. The effect of a city will be felt in the air above it and Figure 11.2 shows some of these influences.

11.4 Windflow

A city influences air currents in both mechanical and thermal ways. The natural changes caused by obstacles being placed in the path of the air will give rise to areas of increased ventilation and areas of relative calm. In addition, wind directions will be changed, in some cases so completely as to bring about an air flow completely opposite to that of the prevailing, 'free air' direction. In other words, turbulence is greatly increased so that eddies and independent circulations are to be expected. An interesting study in New York, comparing the Central Park Observatory and the La Guardia Airport wind speeds, showed a reduction at the city location of about 20%. It has been postulated that, under calm conditions, a city will set up its own wind flow (somewhat comparable to a land/sea breeze), with dT equal to 3 deg C (5·4 deg F). A speed of 3 m s^{-1} (6·6 mph) would then occur on the edge of town.

11.5   Fog and Visibility

As cities and industrialisation increase, so does the frequency of fog. In the environment of the city it is often impossible to distinguish between occasions of fog and those of pollution type 'fog-smog'.

Studies concerned with all large cities, especially those where heating is required, have shown both an increase of the difference in the number of fog-days between urban and rural areas and also in fog frequency with growth of the city. Some data for Paris showed that light fogs were six times more frequent in the city than in the suburbs, while dense fogs (visibility less than 100 m) were nearly twice as frequent.

11.6   Humidity

Cities have lower values than the countryside for both relative and absolute humidity. Detailed studies concerned with humidity are not common, and the seasonal pattern of the difference depends largely on the local precipitation patterns. The W.M.O. conference on Urban Climates (1970) does not have a single paper dealing with this interesting aspect.

11.7   Cloudiness

The more frequent fogs and pollution plus the extra convection and turbulence over the city will combine to lead to an increase in cloudiness. However, the established increases are small (*see* Table 11.1) and their seasonal pattern, as with humidity, is related to the climate of the region. The maximum difference between urban and rural cloud amounts appears in the early morning hours when fogs form, and in the early afternoon as convection develops.

11.8   Precipitation

As with some of the other variables, this is related mainly to air contaminants acting as condensation nuclei. It is well substantiated that there is a city influence giving rise to an increased amount of rainfall, although, because of the air flow patterns, the maximum difference can be displaced away from the city centre. One of the most interesting, and controversial, examples is the La Porte case. La Porte, 53 km (32 miles), downwind of the steel mills in Gary and Chicago has 31% more rain, 38% more thunderstorms and 246% more hail than the neighbouring countryside. Here, at La Porte, the moisture from Lake Michigan converges with the condensation nuclei, heat and moisture from the eastward moving air, and the region experiences a climate modified by man.

Snowfall presents an interesting problem. Normally, the heat from the city will lead to less snow falling there than in the country, but pollution can increase nucleation, and on days with light snowfall more is often reported within the city.

11.9   Air Pollution

11.9.1   Introduction

There has never been a totally unpolluted atmosphere. In the early period of the formation of the Earth, all the gases associated with violent geological activity were being emitted into the atmosphere. Before man came on the scene, volcanoes, decaying flora and fauna, and forest fires were adding their noxious fumes to the air. Air pollution is not as readily definable as many at first appear to be the case. Generally, the definition used by the public encompasses a very wide range. For instance, some pollutants, such as mercaptan from paper mills, can be detected by the nose in small quantities that no instrument can record, while as another example, a 'tar' smell (or a 'fried chicken' aroma) may be considered pleasant by some people and unpleasant by others. If we limit our definition to harmful pollutants, then, not only are we faced with the complication that the criterion of 'harm' is time-dependent, but, we may not be aware that a specific substance can cause 'harm'.

Air pollution, which can occur anywhere, is often thought of as a particularly urban problem, generally because of the concentration of pollution sources in such a region. Historically, it is noted that the mediaeval towns suffered from air pollution by wood smoke, rubbish, sewage, and the tanning industries, among others; in fact, Queen Eleanor left Nottingham, England, in 1257 because of the 'fumes and obnoxious atmosphere', and in 1273 Parliament passed an Act prohibiting the burning of coal. It is recorded that a London man was executed in 1307 for breaking this law. During the sixteenth century a deputation of women came to Westminister to complain to Queen Elizabeth I concerning 'the filthy, dangerous use of coal'.

Air pollution, as a social problem, therefore, dates from the thirteenth century, but a scientific and technical review cannot begin much before about 1870 due to technological limitations. Smoke and soot surveys were made in the 1870s in London, in 1885 in Paris and 1907–12 in St. Louis, Chicago and Pittsburgh. Generally, it was not until the early twentieth century that special state or city controls began to appear—controls that, when enforced, have been instrumental in very dramatic cleaning of polluted media—witness the present conditions in Pittsburgh, London, Manchester and of the River Thames compared with one or two decades ago.

In the realm of disasters due to air pollution, it is sufficient to cite four cases. The earliest was in December 1930 in the Meuse Valley, Belgium, when more than sixty people were killed by polluted air—an event that brought the seriousness of air pollution to the forefront. In October 1948 in Donora, Pennsylvania, during an interval of four days, about fifteen deaths were ascribed to pollution from a steel mill and a zinc reduction plant. At Poza Rica, Mexico, a large, accidental spillage of hydrogen sulphide gas occurred on 24 November, 1950, that caused the death of twenty-two persons. The classic 'disaster' is that of London during 5–9 December, 1952 when an estimated excess of four thousand deaths occurred during a particularly bad 'smog' pollution incident. In all four cases anticyclonic weather conditions prevailed, together with a temperature inversion (*see* Section 11.9.2). In the London episode, the concentrations of sulphur dioxide were of a level that is not normally considered toxic. As Heimann (1961) notes 'it was not known then, nor is it now, exactly what substances in the air should be measured in such episodes as the definite culpable agent'. London has suffered other severe episodes; in February, 1880 (one thousand deaths), January 1956 (one thousand deaths), December, 1957 (seven hundred deaths) and December, 1962 (seven hundred deaths), while New York has had three incidents, each of which caused about two hundred deaths—in November, 1953, January, 1963 and November, 1966.

11.9.2   Sources

There are three basic types of pollution sources:

1. point—such as effluent from a chimney
2. line—for example, along a major highway
3. area—where a number of sources 1 and 2 are intermingled.

In each case the amount of reduction (dilution) that takes place is a function of the meteorological conditions and the manner in which they vary in both space and time for the area of study.

The meteorological variables that have the most impact upon the dispersion of pollutants from their sources are wind speed and the thermal stratification in the low layers of the atmosphere. As we have noted earlier, cool air will have a greater density than warm air. Thus, if a temperature inversion (warm air above) occurs, the air will be stable. There is a critical gradient, or lapse rate, of temperature which if exceeded will produce an unstable air mass—an air mass that when slightly disturbed will begin an eddy or turbulent motion. Therefore, thermal stratification (atmospheric stability) has a great effect on a plume of smoke. The five basic patterns are shown in Figure 11.3.

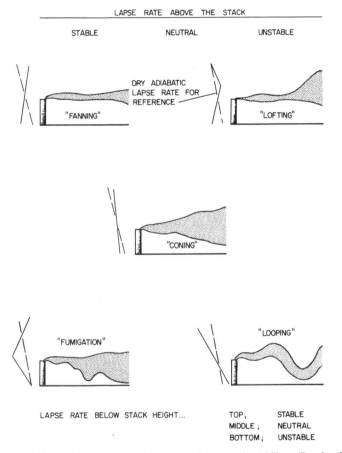

FIGURE 11.3 Plume types and atmospheric thermal stability. (Reprinted by permission of William P. Lowry from *Weather and Life*, Academic Press, 1969)

The horizontal dispersion of a plume is related to the wind speed and its variability. Many mathematical models exist to calculate the mean concentrations of pollutants downwind, although, due to local variants, these are not accurate on the micro-scale.

In the United States, the National Weather Service has devised models that are used to assess whether or not a potential situation exists for air pollution. The criteria include low wind speeds, shallow mixing depths and the continuation of these for at least 36 hours. Precipitation tends to clean the air and rainfall amounts are often entered in the model. It should be noted that even if a stack is emitting only water vapour (hardly a true pollutant) this will tend to alter the atmospheric conditions in the vicinity.

Naturally, the wind speed is an important element, for under calm conditions the pollutants will be restricted to a small volume in the vicinity of the source. Anticyclones, the high pressure cells, generally bring about stable conditions in which air movement is very slight; this is distinctly different from the faster-moving cyclones that often give rise to high wind speeds, and help to clean out pollution.

11.9.3   Pollutants

Industry can feed many pollutants into the air, the major contaminant being sulphur dioxide and its derivatives. However, the odour of mercaptans is usually the most noxious. The major automobile pollutants are carbon dioxide and the oxides of nitrogen. The latter participate in photochemical reactions and lead to the formation of ozone and peroxyacetylnitrate (PAN); both PAN and ozone have effects on plants, animals and humans. In urban areas where dust is greater than in the rural environment, great suffering can result for dust-allergic persons.

Air pollutants fall into many categories—basically the gaseous and the particulate. Various countries and cities have emphasised specific pollutants; for instance, in Great Britain extreme stress has been put on the smokiness or dirtiness of the atmosphere because the main pollutant has been from sulphur. In the United States the National Air Sampling Network has tended to specialise on particulate measurements. Generally, pollutants occur in very small quantities and verification of mathematical models is expensive and must be conducted over long periods to cover the various synoptic situations.

References

Chandler, T. J. (1965) *The Climate of London*, Hutchinson, London, 292 pp.

Griffiths, J. F. and Griffiths, M. J. (1974) *Bibliography of the Urban Modification of the Atmospheric and Hydrologic Environment*, N.O.A.A. Tech. Mem. Ed. 21, 92 pp.

Heimann, H. (1961) 'Effects of air pollution on human health' in *Air Pollution*, W.H.O., Geneva, 159–228

Landsberg, H. E. (1970) 'Climates and urban planning', in *Urban*

Climates, W.M.O. Tech. Note No. 108, World Meteorol. Org., Geneva, 364–72

Lowry, W. P. (1967) 'The climate of cities', *Sci. Am.*, **217,** 15–23

Sundborg, A. (1951) Climatological studies in Uppsala, Uppsala University Geografiska Inst., *Geographica*, **22**

Terjung, W. M. and Louie, S.S-F. (1973) 'Solar radiation and urban heat islands', *Ann. Assoc. Am. Geographers*, **63**(2), 181–207

World Meteorological Organization (1970) *Urban Climates*, W.M.O. No 254, Tech. Publ. 141, 390 pp.

Suggested Reading

American Meteorological Society (1972) *Preprints of Conference on Urban Environment and Second Conference on Biometeorology*, 317 pp.

Kratzer, A. (1962) *The Climate of Cities*, U.S. Dept. Commerce, Off. Tech. Services, AFCRL 62–837, 221 pp.

Landsberg, H. E. (1972) 'Climate of the urban biosphere,' *Biometeorology*, **5,** II, 71–83.

Landsberg, H. E. (1969) *Weather and Health*, Doubleday, New York, 148 pp.

U.S. Dept. Health, Education and Welfare (1962) *Air Over Cities*, Public Health Service, Robert A. Taft Sanitary Engineering Centre, Tech. Report A 62–5, 290 pp.

12

Miscellaneous Applications

12.1 Introduction

Weather and climate play a role in so many of Man's activities that it is difficult to encompass all fields. However, this chapter will endeavour to give a cross-section of the broad spectrum of applications but, of necessity in such a small book, some facets will be omitted. For example, weather has influenced the course of history, as the reader of *Climate, Man and History* (Claiborne, 1970) will soon learn; for instance, with regard to the Vikings giving '. . . the only unambiguous case of a major historical event set off by climatic change', the cooling of the Scandinavian area by a few degrees reduced drastically the farming area, leading to the necessity of increasing overseas forays and plundering for food. In addition, if it had not been for the excellent forecasting before and during the D-day landings in Europe in 1944, history might have been changed radically. Climate has also played an important role in the evolution, the development or the demise of organisms—ranging from the extinction of reptiles such as the land dinosaurs, pterosaurs, ichthyosaurs and plesiosaurs at the end of the Cretaceous period, to the modification of the coats of certain animals during the regular seasonal climatic variations.

12.2 Energy

In these days of the awareness of energy consumption, it is particularly relevant to study the role of climate. We have already seen how building design should take cognisance of the atmospheric conditions but even so, most heavily populated areas of the world do need to use energy for heating and/or cooling purposes in homes, factories, offices and public places. To be aware of the demand, many power plants use the concept of degree days (Section 9.11). For heating requirements, it has been found that the summation of degrees below a chosen threshold gives a close correlation with energy use. The value of the threshold tends to vary, over narrow limits, in different countries but in the USA it is assumed to be 65°F (18°C). Under this condition the heating degree days vary from an average of over 15 000 in central Alaska, through 6000, 5000, 3000 and 1400 at Chicago, New York,

San Francisco and Houston, respectively, to 250 in Miami and 0 in Honolulu.

A similar approach, using the summation of temperature above a selected threshold, is utilised to estimate cooling needs. Naturally, other climatic elements such as humidity, radiation and wind also play a part but temperature is the dominant influence. In the USA 65°F (18°C) is again used as the threshold and values range from 50 per year in central Alaska, through 100, 650, 850 and 3000 at San Francisco, Chicago, New York and Houston, to over 4000 in Miami and Honolulu. In Aberdeen and London, the heating degree days total about 6500 and 5200 per year, respectively, while the cooling days total less than 100.

12.3 Transport

For convenience this section is divided into only four aspects: air, rail, road and water transport.

12.3.1 Air

Climatic knowledge plays a major role in the selection of airport location and for the layout of runways, special studies on visibility, fog, clouds, wind velocity and gustiness must be undertaken. The actual flight effects of weather are too numerous to list but, as is well known, the 'weather briefing' is an essential part of every pre-flight preparation. A special application is in the use of runway temperatures for the calculation of freight loads, an aspect dependent upon the changes of the density of the air which in turn affects the lift.

A survey of US General Aviation aircraft accidents in 1966 showed that while only 15·6% of all accidents were weather-related, for fatal accidents the percentage increased to 32·6%.

12.3.2 Rail

A study by Hay (1957) listed six major causes for the interruption of railway services during the winter, with over 90% related to snow or ice hazards. In Table 12.1 the weather effects on rail traffic are listed and described.

12.3.3 Road

Climatology can aid greatly in the problem of highway design, layout and construction for, as with airports, the need to minimise the incidence of fogs, icing, drifting snow and slick surfaces is obvious. Studies in Canada (Maunder, 1970) and in Australia (Robinson, 1965) have shown that there is a significant increase in road accidents on rainy days.

TABLE 12.1

Weather effects on railroad operations (Hay, 1957)

Climatic Variable	Effect
1. Roadway and Tracks	
Large temperature variation	Excessive rail expansion.
	'Sun kinks' develop—bend rails, keep drawbridges from closing.
Dry weather	Increases danger from fires.
Low temperature	Frost heaving of track and roadbed.
	Ice and freezing may damage bridge piers.
	Steel rails become brittle.
	Switches may freeze.
Snow	Switches, joints, crossings, etc. can become impacted and rendered inoperative.
	Drifting or slides can block open tracks especially in defiles.
Wind	Significant when causing snow or sand drifts of lower visibility.
	Windchill effect.
	Special problems with hurricanes and tornadoes.
Moisture	Excessive humidity can cause rusting and corrosion.
	High rainfall can cause soft spots in roadway or slippage of subgrade materials, landslides.
	Floods and washouts can occur.
2. Signals and Communications	
Large temperature variation	Breaking of wires with repeated expansion and contraction.
Low temperature	Moving parts can freeze.
Snow	Signals covered by drifting snow.
Snow slides	Carry away signals or wires.
Ice	Downing of lines and poles.
3. Yards and Terminals	
Temperature	Cars roll freer in dump yards with high temperatures, more slowly with low temperatures—may require two dumps.
Low temperature	Coal may freeze.
	Diesel fuel oil needs heating when below −18°C.
Snow	Heavy falls as well as drifting can disrupt yard.
4. Locomotives and Rolling Stock	
High temperature	Increased incidence of hot boxes in cars.
	Perishable materials in shipment must be handled in refrigerator cars.
	Domes needed in liquid cargo cars so liquids can expand and gases can collect and escape.

TABLE 12.1 (*continued*)

Low temperature	Frost or ice on rails causes loss of adhesion—coefficient of friction of 0·15 for slippery rails *v*. 0·25 for dry rails would mean 40–50% reduction in traction and in hauling capacity. Diesels likely to freeze in very cold temperatures should be housed at −40°C or below. Lubrication less effective, resulting in overheated journals and hotboxes.
Wind	Blowing sand may block tracks, clog air filters, increase wear on bearings.
Rainfall	Maximum depth of only 75 mm over rails is permitted for diesels in order to keep moisture out of motors.

Freezing and thawing cycles can play havoc with road surfaces and many engineers use the concepts of freezing degree days, freezing index and thawing index (Section 12.6) to assist in decision-making. Heavy rains are also a problem in many areas, especially the tropics, and it remains to be seen whether the Trans-Amazonia Highway will stand the impact of the weather over many years.

12.3.4   Water

In the days of sailing ships, and even earlier in the times of manpowered galleys, the climatic patterns and weather vagaries played an important role in water transport. To a degree, this impact has now been reduced, due to good forecasts of wind velocity and resultant wave patterns, temperature and fog incidence. Ship weather-routing, discussed in White (1971) and Kruhl (1971), is concerned with the aspects of time saving and the smoothness of the voyage—to lessen possible hazards to cargo, ships and passengers—and very considerable savings of from 10 to 20 hours in a trans-Atlantic winter crossing are cited as averages. The temperature is important because of its direct concern to the shipping (ice floes, ice bergs, ice formation) and also due to its effect upon conditions within the hold for, in a hold that is not thermally controlled, the upper layers will be close to the air temperature while the lower layers will approach the sea temperature. Temperature changes during a voyage can give serious condensation problems, especially where potentially corrosive materials (such as automobiles) are being transported.

12.4 Communications

Perhaps the most fundamental impact of climate and weather upon communication is in the effect on radio reception. Changes in atmospheric conditions lead to problems, such as static, and cause wide variation in the refraction of the radio waves, with resultant fluctuations in the received signal.

Overland cables are subject to many climatic stresses (icing, high winds, lightning) while temperature changes will give rise to expansion and contraction of the wires. The use of water-proofed underground cables would make the system almost entirely free from weather influences, but the initial cost is extremely high, compared with the installation of the regular overhead cabling.

12.5 Tourism

In recent years, many countries or regions have found that the tourist industry has become a major item in their economy. While climate cannot be considered to be the only factor, it is true, nevertheless, that a benign climate can attract the tourist. There are plenty of examples of this statement—the popularity of the Mediterranean area with the peoples of northern and north-western Europe, the flow of the 'fortunates' (snow birds) to Florida, the Caribbean and Hawaii during the long winter of the northern USA, and even, in a smaller dimension, the attractiveness of a winter in a dry region such as Las Vegas (Nevada) compared to Los Angeles (California) only 400 km (250 miles) away. Of course, the winter resorts need ideal skiing and skating conditions, both of which can suffer greatly from the vagaries of the weather.

Although not strictly a tourist activity, some areas have become extremely popular as retirement regions and this is due, in no small measure, to their pleasant climate—generally, that means mild winters and a low to moderate amount of rainfall.

Any outdoor recreation or activity, such as sailing, swimming, bicycling, fishing, is affected by the weather conditions, but it is not a simple task to assess the effect. However, its importance is clear when the 1970 Annual Report of the Meteorological Office (UK) shows that holiday and recreational weather enquiries were the most numerous categories handled.

12.6 Appendix

1. *Freezing degree-days* The number of freezing degree-days for a day with a mean air temperature of $\bar{T}$ is $(\bar{T}-32)$. When this value is

negative, freezing degree-days are generated; when positive they are thawing degree-days.

2. *Freezing index* This is simply the number of degree-days between the highest and lowest points on a curve of cumulative degree-days against time for one freezing season.

3. *Thawing index* This is similar to the freezing index but is measured from the lowest to the highest points. When the freezing index regularly exceeds the thawing index a region of permafrost will occur.

12.7 The Future

A meteorologist's training is generally directed towards forecasting for a period of a few hours to a few days—and not always with complete success. It is, therefore, a rather daring venture to attempt to look ahead for years and decades, but perhaps such an undertaking is worthwhile.

During recent years there has been a shift of emphasis in meteorology, and it now appears that the science is pursuing vigorously three main branches—numerical prediction, satellite studies and practical applications, particularly with regard to agriculture.

The field of numerical prediction is concerned not only with weather forecasting but also with meso- and micro-scale facets, such as diffusion and turbulence models (Chapter 11) and macro-scale studies in global climatic modelling. The latter research is attempting to determine how the atmosphere may react to changes in energy input levels.

In the realm of forecasting we must ask if we are stressing too much the aim of a perfect, short-term weather prognosis. Perhaps a meteorological parallel to the Heisenberg Uncertainty Principle is involved so that, because of imponderable motions and varying turbulence-layer characteristics, the best we can hope for is a probability estimate type of forecast.

Satellite studies are increasing our knowledge of weather conditions, not only through the pictures of cloud systems, but also via sensors in wavelengths other than the visible, to give information on the variations of temperature over the oceans, and on temperature and humidity patterns with height. The latter is a particularly intriguing development, because knowledge of the upper atmosphere has been gleaned previously from very few soundings over the globe and the new method could yield a much more complete idea of upper tropospheric and stratospheric conditions. Satellite data are also being studied for use in estimating moisture in the top layers of the soil and the moisture stress in vegetation.

The applied field offers almost limitless possibilities, as I trust this little volume has shown, but with the recent emphasis on world food problems, it is in the agricultural sphere that most effort is being concentrated. For many years it has been possible to estimate the time of maturity of certain crops. Now attempts are being made to forecast the yields using meteorological variables in the equations. So far, the attempts have met with varying, but limited, success. It is not only the direct influence of weather and climate on crops that will come under investigation, though, for research into the effects of the atmosphere on diseases and pests is certain to intensify. Some of these, such as the potato blight (Mather, 1974) and outbreaks of the desert locust (W.M.O., 1965) have already been related to meteorological variables; many others will be.

A big problem that has confronted meteorologists recently is that of significant climatic changes. The repercussions of even small fluctuations can be severe (Chapter 4). If, as has been suggested, the climate is likely to become more variable than in recent decades, then we must be prepared for such changes and it is heartening to note the leading role played by meteorologists of the International Federation of Institutes for Advanced Study in a projected investigation on 'The Social, Moral, Legal and Political Impact of a Reliable and Credible Long-Range Climate Prediction System'. Now, as Godske (1969) has remarked, we have reached the stage of 'inspection for inspiration', and it is thought-provoking to note that another of his statements 'synopticians will gain in stature and become climatologists' is already coming true.

Finally, to the subject of weather modification. The weather is never exactly what we would like it to be, there is a longing to 'mould it nearer to the heart's desire', but could we ever agree on just what weather we want tomorrow and what might be the consequences if we could? Let us recall that the feedback mechanisms within the atmosphere are not completely understood. It may be possible to tip a delicate balance in a dangerous manner. A recent science-fiction novel by Ballard (1974) gives food for thought when, in *The Drought*, he writes

> Covering the off-shore waters of the world's oceans, to a distance of about 1000 miles from the coast, was a thin but resilient monomolecular film formed from a complex of saturated long-chain polymers, generated within the sea from the vast quantities of industrial wastes discharged into the ocean basins during the previous fifty years. This tough, oxygen permeable membrane layer on the air-water interface prevented almost all evaporation of surface water into the air space above.

Science-fiction, yes, but we never knew about photochemical smog—until it happened!

It may be a long time until we can reach the dream world of Camelot where '—July and August cannot be too hot, and there's a legal limit to the snow here—The winter is forbidden till December, and exits March the second on the dot, by order summer lingers through September.'

References

Ballard, J. G. (1974) *The Drought*, Penguin Books, Harmondsworth, England

Claiborne, R. (1970) *Climate, Man and History*, Norton and Co., New York, 444 pp.

Godske, C. L. (1969) 'The future of meteorological data analysis,' in *Data Processing for Climatological Purposes*, W.M.O., Tech. Note No. 100, 52–63

Hay, W. M. (1957) 'Effects of weather on railroad operation, maintenance and construction in industrial operation under extremes of weather', *Meteorol. Monographs*, 2(9), 10–36

Kruhl, H. (1971) 'Ship-routing activities in the Seewetteramt, Hamburg', *Mar. Obs.*, 41(231), 27–9

Mather, J. R. (1974) *Climatology, Fundamentals and Applications*, McGraw-Hill, New York, 412 pp.

Maunder, W. J. (1970) *The Value of Weather*, Methuen and Co., London, 388 pp.

Robinson, A. H. O. (1965) 'Road weather alerts. What is weather worth?' Australian Bureau of Meteorology, Melbourne, 41–3

White, G. A. (1971) 'Practical and economic aspects of ship routing', *Mar. Obs.* 41(231), 25–6

World Meteorological Organization (1965) *Meteorology and the Desert Locust*, Tech. Note No. 69, 310 pp.

Suggested Reading

Maunder, W. J. (1973) 'Weekly weather and economic activities on a national scale: an example using United States retail trade data'. *Weather*, 28(1), 2–18

Perry, A. M. (1972) 'Weather, climate and tourism', *Weather*, 27(5), 199–203

Index